The Organization and Efficiency of Solid Waste Collection

E. S. Savas
Columbia University

With contributions by

Daniel Baumol
Franklin R. Edwards
Frank P. Grad
Bennett C. Jaffee
Christopher Niemczewski
Barbara J. Stevens
William A. Wells

Lexington Books
D.C. Heath and Company
Lexington, Massachusetts
Toronto

Andrew S. Thomas Memorial Library
MORRIS HARVEY COLLEGE, CHARLESTON, W.VA.
97413

352.6
Sa93o

Library of Congress Cataloging in Publication Data

Savas, E.S.
The organization and efficiency of solid waste collection.

Includes bibliographical references and index.
1. Refuse and refuse collection. 2. Refuse and refuse disposal. I. Title.
HD4482.S38   352'.6   76-43606
ISBN 0-669-01095-2

Copyright © 1977 by The Trustees of Columbia University

Reproduction in whole or in part permitted for any purpose of the United States Government, including reproduction by National Technical Information Service in microfiche only.

All rights reserved. No part of this publication may be reproduced or transmitted in any form or by any means, electronic or mechanical, including photocopy, recording, or any information storage or retrieval system, without permission in writing from the publisher. The provision of the United States Government constitutes the only exception to this provision. This book was prepared with the support of NSF Grant APR74-02061. However, any opinions, findings, conclusions, or recommendations herein are those of the researchers and do not necessarily reflect the views of NSF.

Published simultaneously in Canada.

Printed in the United States of America.

International Standard Book Number: 0-669-01095-2

Library of Congress Catalog Card Number: 76-43606

**The Organization and
Efficiency of Solid
Waste Collection**

To Helen, Jonathan, and Stephen

Contents

	List of Figures	xi
	List of Tables	xiii
	Acknowledgments	xix
Chapter 1	**Introduction**	1
	Solid Waste Management	3
	Definitions	3
	The Solid Waste System	4
	Organization of the Book	9
Chapter 2	**The History of Solid Waste Management** *Christopher Niemczewski*	11
	In the Beginning	11
	Classical Civilization	13
	The Medieval Era	14
	The Fifteenth to Eighteenth Centuries	16
	The Nineteenth and Twentieth Centuries	19
	Conclusion	22
Chapter 3	**The Organization of Solid Waste Collection: A Framework for Analysis**	25
	Service Elements	25
	Service Arrangements	29
	Service Patterns	33
	Summary	33
Chapter 4	**The Organization of Solid Waste Collection: Findings**	35
	Previous Studies	35
	The Columbia University Survey	44
	Survey Findings	45
	Extrapolation of Findings	59
	Summary	63
Chapter 5	**Service Levels for Residential Refuse Collection**	67
	Frequency of Service	67

	Pickup Location	67
	Excessive Service Levels	72
Chapter 6	**Financing Solid Waste Collection** *E.S. Savas, Daniel Baumol,* and *William A. Wells*	79
	Current Practices for Financing Solid Waste Collection	79
	Mandatory Refuse Collection	84
	The Effect of Payment Mode on the Cost of Refuse Collection	85
	User Charges and the Cost of Collection	91
	Summary	95
Chapter 7	**The Cost of Residential Refuse Collection** *Barbara J. Stevens*	97
	Introduction	97
	Cost Components in Residential Refuse Collection	98
	Determinants of the Cost of Refuse Collection	107
	Conclusions	116
Chapter 8	**Service Arrangement and the Cost of Residential Refuse Collection** *Barbara J. Stevens*	121
	Introduction	121
	Cost and Organizational Arrangements	127
	Conclusions	135
Chapter 9	**Local Government Regulation of Residential Refuse Collection by Private Firms** *Franklin R. Edwards* and *Barbara J. Stevens*	139
	Introduction	139
	The Case for Government Intervention	141
	The Efficiency of Alternative Regulatory Schemes	145
	Summary	149
Chapter 10	**Contracts for Residential Refuse Collection** *Bennett C. Jaffee*	153
	Introduction	153
	General Requirements of the Contract Document	153
	Discussion of Findings	157
	Policy Guidelines	167

Chapter 11	**The Role of the Federal and State Governments** *Frank P. Grad*	169
	Introduction	169
	The Impact of Federal Legislation	170
	The Growing Role of State Governments in Solid Waste Collection	173
	State Legislation on Collection Since 1965	176
	Summary	182
Chapter 12	**Policy Conclusions and Recommendations**	199
	Appendix A: Universal Telephone Survey	205
	Appendix B: Determining the Cost of Municipal Collection	223
	Appendix C: Determining the Prices of Private Firms	255
	Index	279
	List of Contributors	287
	About the Author	289

List of Figures

1-1	Flow Diagram of the Solid Waste Generation, Collection, and Disposal System	6
4-1	Arrangements Used for Residential Mixed Refuse Collection by City Size	57
5-1	The Frequency of Collection by Arrangement, for Cities of Different Sizes	75
5-2	The Frequency of Collection by Arrangement, for Cities in Different Population Densities	76
5-3	The Frequency of Collection by Arrangement, for Cities in Different Regions of the Country	77
6-1	Distribution of the Ratio of Billing Costs to Total Cost, for 39 Municipal Solid Waste Collection Systems	88
6-2	Total Cost of Refuse Collection Varies by Mode of Payment and Family Income because of Deductibility of Property Taxes on Federal Income Tax Returns	92
6-3	User Fees vs. Collection Costs for Cities with Municipal Collection	94
6-4	User Fees vs. Collection Costs for Cities with Contract Collection	95
8-1	Annual Cost per Household for Once-per-Week Curbside Refuse Collection; Estimated Holding Wage Rate, Refuse per Household, Density, Service Level, and Temperature Variation Constant	133

List of Tables

1-1	Classification of Solid Waste	5
1-2	Solid Waste Generation in the United States	7
1-3	Composition of Municipal Solid Waste, 1971	8
3-1	Organizational Elements in Solid Waste Collection	28
3-2	Service Arrangements for Solid Waste Collection	30
3-3	Nomenclature and Definition of Service Arrangements	31
3-4	Definition of Contract, Franchise, and Private Arrangements Where the Service Provider Is a Private Firm	31
3-5	Municipal Service Arrangements in Indianapolis for Collection of Mixed Residential Refuse	32
3-6	Pattern of Municipal Solid Waste Collection Services for a City	34
4-1	Surveys of Organizational Arrangements for Refuse Collection	36
4-2	Partial Presentation of Data from 1974 ICMA Survey	42
4-3	Summary of Survey Findings	43
4-4	Number and Percentage of Cities, by Service Provider, for each Type of Service Recipient	46
4-5	Comparison of Survey Findings	48
4-6	Percentage of Population Living in Cities That Have the Indicated Service Provider, by Type of Service Recipient	48

4-7	Percentage of Population Served by Indicated Service Provider, for Collection of Residential Mixed Refuse from Other Than Multiple Dwellings	49
4-8	Observed Service Arrangements for Solid Waste Collection, for All Service Recipients	50
4-9	Number and Percentage of Cities, by Arrangement, for Each Type of Service Recipient	51
4-10	Number and Percentage of Cities in Which a Class of Service Recipient Is Served by Only a Single Arrangement	53
4-11	Distribution of Cities by Arrangement and Combinations of Arrangements for Collection of Mixed Refuse from Residences Other than Multiple Dwellings	54
4-12	Comparison of Survey Results for Cities Larger than 5,000 in Population	55
4-13	Service Arrangements for Collection of Residential Mixed Refuse	56
4-14	Arrangements for Other Types of Residential Service	58
4-15	Changes in Residential Service Arrangements, in Terms of Service Providers	60
4-16	Service Arrangements for Large Cities	61
4-17	Service Arrangements by Service Recipient for Suburban Cities in SMSAs with the Ten Largest Cities in the Sample	62
4-18	Logic of Extrapolation of Findings	63
5-1	Residential Service Level: Frequency of Collection	68
5-2	Residential Service Level: Location of Pickup	70

5-3	Location of Pickup by Mean Annual Family Income, in Number and Percentage of Cities	72
5-4	Frequency of Residential Collection, by Arrangement	74
6-1	Number and Percentage of Collection Systems Using Each Payment Mode, by Service Recipient	80
6-2	Percentage of Residences Using Each Payment Mode	81
6-3	Mode of Payment for Refuse Collection from Other-than-Multiple Dwellings	82
6-4	Payment Mode by Arrangement, for Residential Refuse Collection	83
6-5	Relationship between Mode of Payment and Amount of Refuse Generated	85
6-6	Relationship between Mode of Payment and Frequency of Collection	86
6-7	Relationship between Mode of Payment and Pickup Location	87
6-8	Estimation of Aftertax Cost of a Refuse Collection Service That Is Financed by Property Taxes	90
6-9	Summary of Effect of Payment Mode on the Aftertax Cost per Household for Refuse Collection	93
7-1	Sample of Cities Used for Analysis of Costs	100
7-2	Percentage of Sample from United States Regions by Arrangement Type	101
7-3	Collection Technology, Wage Rate, and Arrangement	102
7-4	Per Household Costs of Residential Refuse Collection, in Cities with Municipal Collection in Dollars per Year	103

7-5	Per Household Costs of Residential Refuse Collection, in Cities with Municipal Collection, by Region in Dollars per Year	105
7-6	Per Household Cost of Residential Refuse Collection in Cities with Municipal Collection, by City Size in Dollars per Year	106
7-7	Mean Values of Efficiency Measures and Cost Determinants	112
7-8	Cost and Cost Determinants	115
8-1	Average Cost per Cubic Yard and per Ton for Collection	128
8-2	Average Cost per Cubic Yard and per Ton for Collection in Cities with Once-a-Week Curbside Collection	129
8-3	City Size and Cost of Refuse Collection	132
8-4	Predicted Annual Cost per Household for Once-a-Week Curbside Collection of Refuse	134
8-5	Management Factors in Refuse Collection	135
9-1	Number of Cities in the Sample, by Institutional and Regulatory Classification	146
9-2	Differences in Average Collection Prices: Effects of Billing Costs, Scale, and Contiguity	149
10-1	Frequency Distribution of Model Contract Terms in Contract Sample	158
10-2	Terms Included in Sample of Contracts	160
10-3	Distribution of Contracts by Contract Provision	162
10-4	Additional Contract Information from City Questionnaires	163

10-5	Prevalence of Terms in Sample Contracts	164
10-6	The Difference between Contract and Franchise Agreements with Respect to Contractual Terms	164
10-7	Effect of City Size on Contract Terms	165
10-8	Effect of City Size on Insurance Amounts	166
10-9	Effect of City Size on Manner of Selecting Private Firm	166
10-10	Term of Contract or Franchise (in years) by City Size	167
10-11	Percentage of Cities That Include Provisions for Adjusting Rate Schedule, by City Size	167
11A-1	State Legislation and Regulation for Solid Waste Management	196
A-1	200 Eligible SMSAs by Region and State	206
A-2	Noncounty Local Jurisdictions in U.S.	212
A-3	Noncounty Local Jurisdictions	213
B-1	Site Selection for SMSA Visits	224
B-2	SMSAs Selected for Site Visits, by Region	225
B-3	Sample Universe of Cities with Municipal Collection, by Region and Population	229
B-4	Municipal Sample by Region and Population Class	230
C-1	Distribution of Cities Initially Selected for Analysis	256
C-2	Selection of Cities with Private Service Providers, for the Economic Analysis	257
C-3	Distribution of Cities Analyzed	258

Acknowledgments

This is a study of cities, or, more precisely, of how cities organize to provide an essential service to their citizens. Thus, it is fitting first of all to acknowledge the contributions of the innumerable city officials and private refuse haulers throughout the nation who patiently withstood endless questioning and provided the insights and suggestions that made this book possible. It is to them that this study is ultimately directed.

This volume results from a large, complex, and lengthy project which was skillfully and successfully supervised on a day-by-day basis by Mrs. Eileen Brettler Berenyi. She deserves the lion's share of the credit for holding the entire project together, of resolving innumerable crises, and maintaining progress in the face of diversity. Besides her organizational talent and personal enthusiasm, her expert research, analytical skills, and cool judgment were continuously called into service.

The authors of the different chapters played a far larger role than their by-lines reveal; in particular, Barbara Stevens contributed greatly to all aspects of the research reported here, and joined me as coprincipal investigator during the second half of the project whose results are reported here. Daniel Baumol analyzed, computed, and managed a large and growing data base, while retaining a broad perspective and providing helpful insights on a broad range of issues. Frank Edwards and Frank Grad are owed many thanks.

I am grateful also to the staffs of the International City Management Association and Public Technology, Inc., for their key assistance, especially in the arduous tasks of obtaining the data that underlies the analyses presented here. Clyde Nunn of the Center for Policy Research, Stephen Shay, Harry Coates, Narendra Thadani, and numerous others at Columbia University participated directly in this effort.

A distinguished advisory panel from the public and private sectors proved invaluable to this effort; although they are too numerous to mention individually, I owe them each a special debt of gratitude.

Finally, when all the theorizing, analyzing, and philosophizing is done, there still remains the work of producing the tangible evidence of all that ratiocination. That job fell to Mrs. Daiselle Crawford, aided by Mrs. Blanche Winikoff, who carried it out with her natural competence, grace, and good nature.

Finally, I am grateful for the support of the National Science Foundation, Research Applied to National Needs division, through its Advanced Productivity Research and Technology program. Our program manager, Dr. Frank Scioli, has provided encouragement and valuable guidance at key points; however, the opinions, findings, conclusions, and recommendations are those of the authors and do not necessarily reflect the views of NSF.

**The Organization and
Efficiency of Solid
Waste Collection**

1 Introduction

Government is big business in the United States. There are almost 80,000 governments in the country, and together they collect revenues which amount to one-third of the gross national product.[1]

Contrary to popular impression, however, state and local governments in the aggregate are much larger than the federal government, by two important measures: (1) Most of the money spent by government for goods and services is spent by state and local governments, not the federal government; the former spent $192 billion in 1974, compared to $116 billion spent by the federal government.[2] (2) State and local government employees number 11.6 million, almost one-fifth of the civilian work force and four times as large as the federal nonmilitary work force. Between 1954 and 1974 their numbers more than doubled—but the payroll increased more than sevenfold.

Given these facts, it is not surprising that productivity has emerged as a dominant problem in managing local government. Many Americans feel that government, particularly local government—which is in the unenviable position of being responsible for daily delivery of highly visible services—is inefficient and ineffective: the disparity between inputs to and outputs from local government looms large in their eyes.

These circumstances nurture the growing belief that significant and enduring improvements in the efficiency and effectiveness of local government can be achieved only by recognizing the institutional nature of the basic problems, designing management strategies to overcome them, and building the political support to do so.

One strategic approach to improving the performance of local government deals with the institutional arrangement by which public services are delivered. There are several different organizational arrangements by which municipalities assure the provision of services to their citizens:

1. providing the service directly, using its employees;
2. contracting with a private organization;
3. contracting with another governmental jurisdiction;
4. issuing vouchers that entitle residents to obtain the service from any authorized supplier;
5. enabling residents to purchase the service from a supplier, by providing a government subsidy to the supplier;
6. leaving it entirely up to each resident to buy the service from any authorized supplier.

Given these alternatives, some fundamental policy issues may be raised: Under what structural arrangement can municipal services best be provided? Must public services be provided by public agencies using public employees and acting as public monopolies? Can competition—public or private—be introduced, and is it advantageous to do so? Under what circumstances, if any, can the private sector provide public services more efficiently and effectively than the public sector? Can the public interest be protected if vital public services are provided by the private sector? How?

A study by the Advisory Commission on Intergovernmental Relations showed that 61 percent of the 2,248 reporting cities contracted for one or more municipal services; contracts were with governments and with private firms.[3] These findings are generally corroborated in a review for the National Commission on Productivity.[4] Health services, social services, and education are provided under contract, as are installation and maintenance of street lights and traffic lights, operation of animal pounds, water supply, street cleaning and maintenance, refuse collection and disposal, snow removal, tax assessment and collection, vehicle maintenance, inspection services of various kinds, emergency ambulance service, and police services, to name a few. Not as well known is the fact that fire protection is purchased by some communities from private firms, reportedly at a lower cost than cities can provide it.[5] In Minneapolis, a major civic effort is under way to examine systematically the public services that might better be provided by the private sector[6] and in California, the widespread "Lakewood Plan" has created a competitive marketplace for municipal services.[7,8] Interest in this entire area of alternative service arrangements can be intense, as evidenced by the reaction to a report comparing the efficiency of public and private service in one large city.[9]

The monopolistic behavior of government bureaucracies has been noted.[10,11,12] "Reprivatization" has been examined and advocated[13,14] and strategies for introducing competition—public and private—in the delivery of municipal services have been analyzed.[15] Nevertheless, despite this attention, there has been little in the way of definitive research in this area. There have been many particular studies of individual services in specific localities, of course, and much in the way of polemic tracts and unsubstantiated assertions, but the basic policy questions remain unanswered:

1. How does the organizational structure of municipal services affect the efficiency and effectiveness of those services?
2. How do publicly and privately provided services compare?
3. What accounts for any differences?
4. What policies and actions in this area should be followed by governments and by private industry?

These issues, so basic to the management of public services, are explored here in the context of solid waste management.

Solid Waste Management

Because of their vital role with respect to public health, solid waste collection and disposal are a concern of government. Generally speaking, they are a responsibility of local government, although states and the federal government are paying increasing attention to these areas and establishing standards through legislative action. While waste collection and disposal do not share the life-saving characteristics of the local emergency services—police, fire, and ambulance—they have even higher political visibility because garbage requires conscious action every day by every family, and if service is unsatisfactory the fact is quickly evident. The importance of this subject to local government is demonstrated by a 1974 survey, in which elected local officials named solid waste management most frequently as a problem, ranking it ahead of even crime and housing.[16]

But in addition to the active involvement by local government in solid waste management, and the relatively high incidence of waste collection by municipal agencies, the private sector, too, plays a very significant role in this service. For these reasons, namely, a vital service, high visibility, and extensive use of public and private services, as well as the *relative* ease of measurement (at least by comparison to police and education services, for instance), solid waste collection affords an excellent opportunity to study the fundamental policy issues identified here.

Definitions

Two kinds of definitions exist for solid waste; one is a conceptual definition, the other a descriptive one, which defines solid waste by listing the kinds of materials that comprise it.

Conceptual Definition of Solid Waste

The conceptual, and widely accepted, definition of solid waste is provided by the American Public Works Association.[17] Wastes are defined as "useless, unwanted or discarded materials," and include solids, liquids, and gases. Those wastes which are solid are referred to as solid wastes or refuse. (The two terms are used synonymously.)

It is symptomatic of the primitive state of analysis of the subject that this commonly accepted definition is quite inadequate. To begin with, uselessness, like beauty, is in the eye of the beholder. The livelihood of more than one million people is dependent upon such "useless material," and, increasingly, useful material and energy are being recovered from it. Furthermore, much useless material is not discarded, but remains in attics, garages, and generally underfoot; that is, not in the solid waste stream. Indeed, according to the definition, a typical suburban garage sale could be described somewhat whimsically as a redistribution of solid waste.

As for being "unwanted," it should be noted that parents are forever discarding their children's most prized and wanted possessions, albeit after contemptuously calling it "junk." In counterpoint, much unwanted material remains undiscarded, simply because the "cost" (usually the effort) of disposal exceeds the "cost" (usually effortless) of retention.

In light of these problems, the following conceptual definition of solid waste is offered: *Solid material which is discarded.* This simple definition ignores the irrelevant issue of the usefulness, value, or desirability of the matter in question, but inasmuch as discarding is an intentional act, it implies that the discarder judges the material to be of *relatively* little current value to *him*.

Descriptive Definition of Solid Waste

Solid waste consists of discarded solid materials resulting from domestic and community activities and from industrial, commercial, and agricultural operations. It does not include solids or dissolved materials in domestic sewage or other pollutants in waterways, such as silt or dissolved matter. Table 1-1 describes and classifies the various kinds of solid waste and also categorizes them by the source of generation and by composition.

The Solid Waste System

Generation and Disposal

A flow diagram of the solid waste system is shown in Figure 1-1. As indicated, solid wastes are generated as a result of both the production of raw materials and goods, and the consumption of those goods.

Of the solid waste generated each year, very little is collected for subsequent disposal. A disproportionate share of the refuse that is not collected consists of agricultural and mining wastes that are deposited in the vicinity of the sites at which they were generated. Other wastes that never enter the collection system are (1) discharged into waterways; (2) incinerated at the site of generation; and (3) discarded as litter that is never collected.

The relatively small fraction of solid waste that is collected consists of municipal waste (including litter), industrial waste, construction and demolition waste, and sewerage sludge.

Much of the refuse collected is disposed of at open dumps, a practice that is being phased out. The remainder goes to sanitary landfill sites and, to a limited extent, to incinerators. The amount that is recycled is negligibly small, at present, in terms of weight.

Table 1-1
Classification of Solid Waste

Type	Composition (Descriptive)	Source
1. Agricultural waste		
(a) crop residues	harvesting residue, vineyard and orchard prunings, greenhouse wastes	farms
(b) animal	manure, slaughterhouse wastes	farms, feedlots, slaughterhouses
2. Mineral waste	earth and rock from mining, extraction, and refining	mines, ore-processing and mineral-refining plants
3. Municipal solid waste		
(a) garbage	waste from the storage, handling sale, preparation, cooking, and serving of food	households, institutions, and commercial establishments
(b) rubbish (or trash)		
(i) combustible (mostly organic)	paper, cardboard, wood, plastics, rags, cloth, leather, rubber, yard waste (grass, leaves)	same
(ii) noncombustible (mostly inorganic)	metal, cans, metal foil, dirt, stones, crockery, ceramics, glass, bottles	same
(c) ashes	residue from fires used for cooking and for space heating	same
(d) bulky waste	stoves, refrigerators, heaters, and other large appliances; furniture, crates, tires, auto parts, tree limbs	same
(e) other municipal waste	street and alley sweepings, catch-basin dirt, contents of litter receptacles in public places, refuse from parks and beaches, dead animals, tree and landscaping refuse (other than yard waste)	streets and other public property
4. Abandoned vehicles	automobiles and trucks	same
5. Industrial waste	waste from industrial processes, manufacturing and power generation including cinders, ash, and scraps and shavings of wood, metals	factories, industrial plants, power plants
6. Construction and demolition waste	lumber, concrete, plaster, roofing, pipe, brick, conduit, sheathing, wire, insulation	construction sites
7. Hazardous waste	pathological waste, explosives, radioactive material, poisons, hazardous chemicals, pesticides	industry and institutions
8. Sewage treatment residues	screenings, grit, digested and dewatered sludge	sewage treatment plants

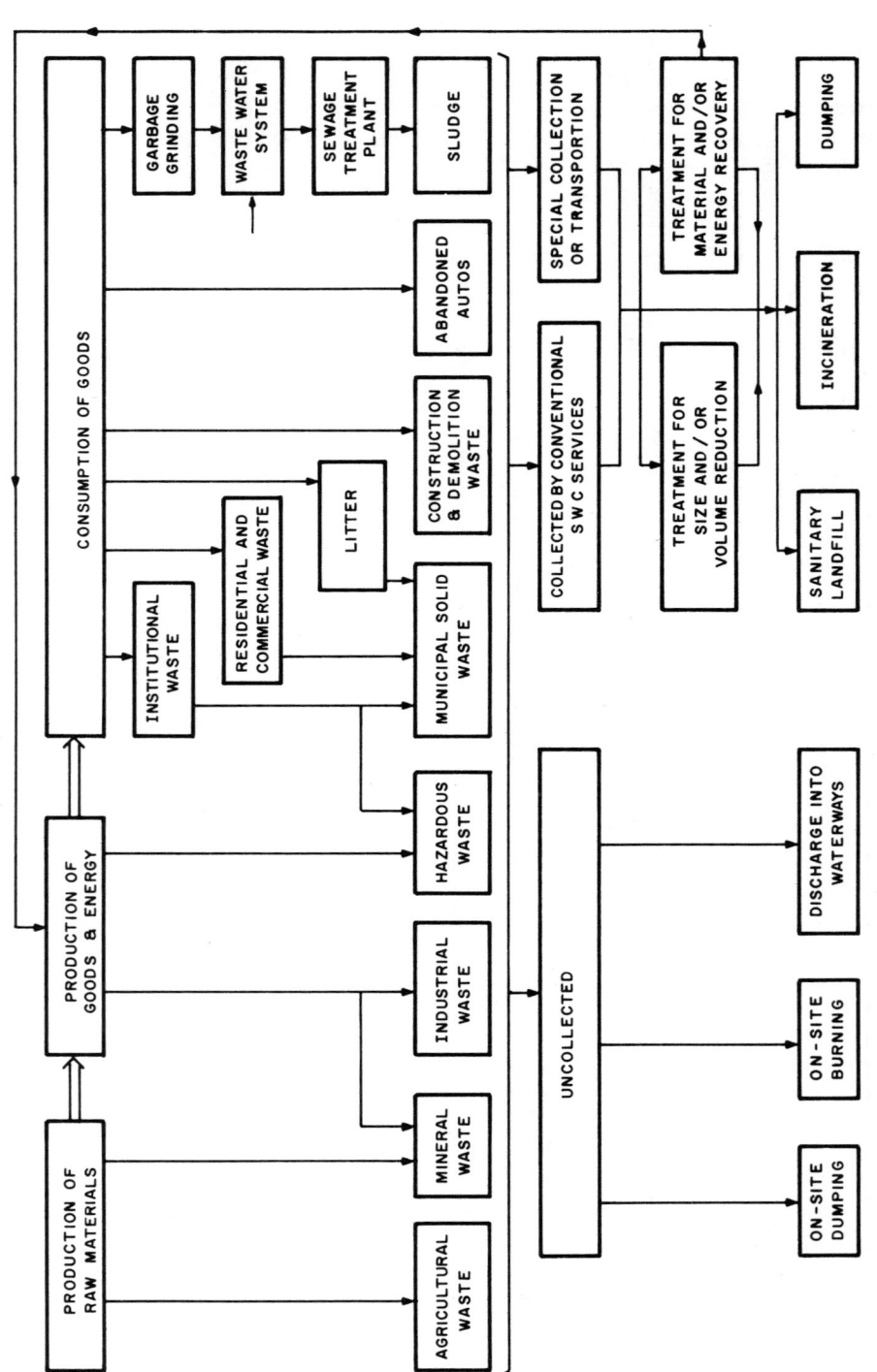

Figure 1-1. Flow Diagram of the Solid Waste Generation, Collection, and Disposal System.

Amounts of Solid Waste

Estimates of the amounts of solid waste generated annually, by source, are confusing, inconsistent, and contradictory. Only rough approximations are available, and these are offered to provide a sense of the order of magnitude of solid waste management activities.

It has been estimated that somewhere in the vicinity of four billion tons of solid waste are generated annually in the United States. Of that, about 90 percent is agricultural and mineral waste, which is not subject to collection. Using the classification scheme of Table 1-1, Table 1-2 presents data on the amount of solid waste, by type, generated in the United States.

Table 1-2
Solid Waste Generation in the United States

Type of Solid Waste	Millions of Tons per Year	Lbs/Person per Day
Agricultural[a]	2,100-2,300	55.6-60.9
Mineral[a]	1,100-1,700	29.1-45.0
Municipal[b]	127-186	3.36-4.92
Residential	90.3	2.39
	(87-102)	(2.3-2.7)
Garbage	17.8	0.47
Rubbish (excluding yard waste)	48.0	1.27
Yard waste	18.1	0.48
Bulky waste	6.4	0.17
Commercial and institutional	34.7	0.92
	(34.62)	(0.9-1.7)
Other municipal waste	6-22	0.17-0.61
Abandoned vehicles[c]	1.0	.03
Industrial	34-105	0.95-2.90
Construction and demolition	5-19	0.13-0.53
Hazardous	N.A.	N.A.
Sewage treatment residues[c]	22	0.61

Source: F.A. Smith, *Comparative Estimates of Post-Consumer Solid Waste*, SW-148 (Washington: U.S. Environmental Protection Agency, 1975).

Notes: Unless otherwise indicated, all estimates are from Smith (1975).
N.A. means not available.

[a]Lower estimates are from Council of State Governments, *The States' Role in Solid Waste Management: A Task Force Report* (Lexington, Ky., 1973). Upper estimates are from U.S. Council on Environmental Quality, *Environmental Quality: First Annual Report* (Washington: Government Printing Office, 1970).

[b]Data for all components of municipal solid waste are for 1971.

[c]Anton J. Muhich, Albert J. Klee, and Paul W. Britton, *1968 National Survey of Community Solid Waste Practices: Preliminary Data Analysis*, Public Health Service publication 1867, Solid Waste Management series SW-3s (Cincinnati: U.S Department of Health, Education, and Welfare, 1968).

Various estimates of residential solid waste generation appear to be reasonably consistent, at about 90 million tons annually, or 2.4 pounds per person per day, in 1971. However, there are wide disparities in the estimates of commercial and institutional solid waste, as shown in Table 1-2.

Composition of Solid Waste

Table 1-3 provides reasonable estimates of the composition of municipal solid waste; it is derived by a combination of input analysis, that is, by computations based on the amount of material used in production, and output analysis, based on the measured composition of collected refuse. These two methods of analysis correlate reasonably well, as reported by the National Center for Resource Recovery[18] and by Smith.[19]

Table 1-3
Composition of Municipal Solid Waste, 1971

Component	Millions of Tons	Lbs/Person per Day	Percentage
Organic	99.0	2.622	79.3
Paper	39.0	1.034	31.3
Yard waste	24.1	0.638	19.3
Food waste	22.0	0.582	17.6
Wood	4.6	0.122	3.7
Plastics	4.2	0.111	3.4
Rubber and leather	3.3	0.087	2.6
Textiles	1.8	0.048	1.4
Inorganic	25.8	0.682	20.6
Glass	12.1	0.320	9.7
Metals	11.9	0.314	9.5
Ferrous	10.6	0.281	8.5
Aluminum	0.8	0.021	0.6
Other	0.4	0.011	0.3
Miscellaneous Inorganics	1.8	0.048	1.4
Total	125.0	3.308	100%

Source: F.A. Smith, *Comparative Estimates of Post-Consumer Solid Waste*, SW-148 (Washington: U.S. Environmental Protection Agency, 1975).
Notes: Includes residential, commercial, and institutional waste, but excludes "other municipal waste" as defined in Table 1-1.
Sums do not equal totals because of rounding.

Organization of the Book

The next chapter reviews the history of solid waste management. Chapters 3 and 4 deal with the organization of solid waste collection services, as they apply to residences and to commercial, institutional, and industrial establishments. The remainder of the book concerns residential collection. Chapters 5, 6, and 7 describe service levels, financing patterns, and the costs of collection, respectively. Chapter 8 analyzes the relative efficiency of the alternative service arrangements. The succeeding chapter examines the rationale for government regulation of residential collection and presents empirical findings about the effectiveness of various regulatory practices. Chapter 10 presents an analysis of the contract documents used by local governments and private firms. The next chapter reviews the role of the federal and state governments in solid waste collection. The final chapter presents conclusions and offers policy guidelines.

Notes

1. *Economic Report of the President* (Washington: Government Printing Office, 1975).

2. Ibid.

3. Joseph F. Zimmerman, "Meeting Service Needs through Intergovernmental Agreements," *Municipal Year Book 40* (1973).

4. Herbert Kiesling and Donald Fisk, *Local Government Privatization/Competition Innovations*: Preliminary report (Washington: Urban Institute, 1973).

5. Roger S. Ahlbrandt, *Municipal Fire Protection Services: Comparison of Alternative Organizational Forms*, Sage Professional Papers in Administrative and Policy Studies, no. 03-002 (Beverly Hills, Calif.: Sage Publications, 1973).

6. *Why Not Buy Service?* (Minneapolis: Citizens League, 1972).

7. Robert O. Warren, *Government in Metropolitan Regions: A Reappraisal of Fractionated Political Organizations* (Davis: Institute of Government Affairs, University of California, 1966).

8. Sidney Sonenblum, John J. Kirlin, and John C. Ries, *Providing Municipal Services: The Effects of Alternative Structures* (Los Angeles: Institute of Government and Public Affairs, University of California, 1975).

9. E.S. Savas, "Municipal Monopolies Versus Competition in Delivering Urban Services," in *Improving the Quality of Urban Management*, ed. Willis D. Hawley and David Rogers, Urban Affairs Annual Reviews, vol. 8 (Beverly Hills, Calif.: Sage Publications, 1974).

10. Anthony Downs, *Urban Problems and Prospects* (Chicago: Markham, 1970).

11. William A. Niskanen, *Bureaucracy and Representative Government* (Chicago: Aldine, 1971).

12. E.S. Savas, "Municipal Monopoly," *Harper's Magazine* 243 (December 1971).

13. Peter Ferdinand Drucker, *The Age of Discontinuity: Guidelines to Our Changing Society* (New York: Harper & Row, 1968).

14. Lyle Craig Fitch, "Increasing the Role of the Private Sector in Providing Public Services," in *Improving the Quality of Urban Management.*

15. Savas, "Municipal Monopolies."

16. Raymond Bancroft, *America's Mayors and Councilmen: Their Problems and Frustrations* (Washington: National League of Cities, 1974).

17. *Solid Waste Collection Practice*, 4th ed. (Chicago: American Public Works Association, 1975)

18. National Center for Resource Recovery, *Resource Recovery from Municipal Solid Waste; a State-of-the-Art Study* (Lexington, Mass.: Lexington Books, 1974).

19. Frank A. Smith, *Comparative Estimates of Post-Consumer Solid Waste*, Solid Waste Management series SW-148 (Cincinnati: U.S. Environment Protection Agency, 1975).

2 The History of Solid Waste Management

The importance of municipal solid waste in terms of public policy derives from its presence in urban areas: it is found where people are, and the more people, the more municipal solid waste; furthermore, the greater the population density, the greater the impact of solid waste on public health and on environmental quality, the greater the cost of dealing with it, and the greater the difficulty of disposal. It has always been so. The history of solid waste collection and disposal is intertwined with the history of the city. It was the growth of urban centers which made necessary increasingly elaborate and complex systems to handle solid waste.

In the Beginning

Prior to the Neolithic revolution (c. 10,000 B.C.), when man was a hunter and food gatherer, the sparse populations and nomadic habits permitted natural decomposition on-site to dispose effectively of any wastes. However, as primitive societies gradually shifted to food production, men lived in more permanent settlements with relatively greater population density, and the obtrusive wastes in their midst could no longer be ignored and left to natural processes. At this stage in urban development, these societies were forced to develop at least simple means for dealing with solid waste.

By 3500 B.C., the city of Ur had an average population density (that is, including the temple precinct which was sparsely inhabited, if at all) of at least 65,000 people per square mile. Mumford[1] quotes Wooley that "the sweepings of house floors and the contents of rubbish bins were simply flung into the streets," so steadily that the street levels were gradually raised and from time to time new doors would have to be cut to keep open access to the houses. Practices that were satisfactory in semipermanent, small villages were not necessarily suitable in an urban environment.

In view of these practices, the Indus Valley city of Mahenjo-Daro, founded no later than 2500 B.C., represents a very significant conceptual leap in terms of waste management. In addition to an elaborate and well-designed drainage system, each house had rubbish chutes, at the foot of which were sometimes bins at street level; also, rubbish bins were placed at convenient locations in the streets. From them waste was removed by scavengers who presumably were under some central authority.[2] (The existence of the chutes implies central planning of buildings and is particularly noteworthy in light of the fact that many modern buildings today are not designed properly to facilitate waste removal.)

Although such elaborate systems were unknown elsewhere, an awareness of the solid waste problem was widespread in India. Refuse collection seems to have been, even through the first half of this century, subsumed in the structure of the caste system. That is, members of the lowest subgroup of the untouchables (the lowest caste) variously either herded their pigs through the streets (so that they would eat the refuse) or collected and disposed of the more obvious wastes (dead animals) themselves. As the caste system dates conservatively from 300 B.C. and more probably from c. 1500-1000 B.C., it is quite likely that this general method of removal is approximately as old. A contemporary administrative description of the Mauryan empire of India (fourth century B.C.) mentions both special punishments for dirtying the roads and arrangements for the employment of scavengers for sweeping the main streets in large cities. It is also possible that in Deccan, by this time, there were baskets placed along the road for the collection of garbage, but this is not confirmed.

In the Egyptian town of Herakopolis (c. 2100 B.C.) wastes in the "nonelite" quarter were treated pretty much as of old—organic wastes were either left on the floor or dumped just outside the dwellings, while inorganic wastes were put into nearby depressions, perhaps to level off the ground. In the elite and religious quarters, however, there was a deliberate effort made to remove all wastes, organic and inorganic, to locations outside the living and/or communal areas, which usually meant into the river.[3]

Mosaic law (c. 1300 B.C.) refers specifically to public sanitary practices; everyone was expected to act as his own scavenger—to remove his own refuse and bury it in the earth. Nehemiah, in connection with the rebuilding of Jerusalem, mentions the "refuse gate"—a particular gate outside of which the city's wastes were dumped.[4] The Talmud (which, although it was written c. 300-500 A.D., is the continuation of an older, oral tradition) says that the streets of Jerusalem had to be washed every day—a stringent requirement in an arid land. Ostensibly, these laws had a strictly religious function; however, it is quite likely that, as in the prohibition of consanguinous marriages, health and religious factors are inextricably intermingled.

A Chinese "Record of Institutions" of about the second century B.C. mentions sanitary police charged with the removal of all carcasses, whether of men or animals, and traffic police whose duty, among others, was to organize the sweeping of the streets in the major cities.

In Crete by 1500 B.C. there were specific places set aside for the disposal of certain organic wastes, which were then reused as compost.

However, waste removal was far from universal. Even in ancient Troy, "fair Priam's city," all wastes were either dropped on the floor inside or at best thrown into the streets. However, as Blegen notes, "Sooner or later there must have come a time when the floor became filled with bones and litter ... [cleanup] was accomplished not by sweeping out the offensive accumulation but by [spreading] a good supply of fresh clean clay ... This was repeated time after time (until it became necessary to raise the roof and rebuild the doorway)."[5]

Classical Civilization

The first "municipal dumps" (at least in what we now regard as the Western world) were organized by the Greeks in c. 500 B.C.; and at approximately this time the Council of Athens began to enforce a regulation according to which scavengers were required to dump wastes no less than one mile from the city walls. Then, in 320 B.C., Athens promulgated the first known edict which forbade the throwing of refuse into the streets.

By 300 B.C. there had evolved a system, both in Greece and in the Greek-dominated cities of the eastern Mediterranean, under which the controllers of the town (astynomoi)—whose first duty was the care of roads and bridges—were responsible for waste removal. In smaller cities this was the responsibility of the controllers of the market (agoranomoi). They had to prevent rubbish being thrown into the street and were to have the streets scavenged as often as they saw fit, the expenses being covered by levies on landowners.[6] This system was sufficiently viable to last for eight hundred years, until the general breakdown of civic order amid the barbarian invasions, and perhaps a thousand years longer in Byzantium and the Muslim world.

Roman collection of solid waste was probably better organized than any other prior to the nineteenth century. Nevertheless, they faced, and were unable to overcome, the problem of disposal of a very large accumulation of wastes. This paradox was most evident in the city of Rome itself. Municipal collection of wastes was confined to state-sponsored events, such as removal of bodies of animals and men after the games and cleansing of streets after parades, and to regular cleaning of principal streets (the latter by decree of Caesar, 47 B.C.). Property owners were legally responsible for the cleanliness of streets fronting their property, although this seems to have been generally ignored. However, independent scavengers collected a significant percentage of residential wastes for resale as fertilizer, and the wealthier residents were able to use their own slaves to perform any necessary further removal.

The problem, then, was twofold. First, the portion of wastes that was not collected became very large in absolute terms as the population of the city grew. Second, disposal arrangements were very crude—the city dumps were merely open pits that ringed the city walls. (Cover was occasionally used; for example, in one case, broken, waste clay from a pottery factory was applied.) Efforts were made to induce those disposing of wastes to carry them farther out from the city. For instance, a sign was found saying "Take your refuse farther out or you will be fined," but it is not clear that this was effective.

The cumulative effects of these practices was that Rome became a chronically unhealthy city. Nevertheless, it is not evident that a viable alternative was available; rather, the size of Rome far exceeded the capacity of available waste removal systems.

After the fall of Rome, sanitation in general and solid waste practices in particular tended to decline. However, several factors mitigated this to some degree. The most notable was the substantial deurbanization of the Western world. Rome itself, a city of some 1.25 million people in its prime, had a population of only about 20,000 in the thirteenth century and 200,000 as late as the sixteenth century. The reduced population density, therefore, rendered traditional methods of waste disposal (tossing it out of the house) more viable.

It is also worth noting that this decline took place almost solely in the Western world. The great Islamic cities (which became, through Byzantium, the heirs to Roman culture and technology)[7] compared favorably as regards cleanliness with any civilization heretofore. They were apparently aware that "sanitation required a system of intelligently spread and scrupulously maintained generous supplies of water to keep the city clean, and, some system of canalization or collection to remove wastes,"[8] hence the need for civic rubbish heaps. Furthermore, inhabitants were subject to almost daily inspection should they be remiss in any matter contingent on water or sanitation and merchants could be fined for not keeping the streets fronting their shops well watered and free of dust. All civic functions were interpreted as religious functions and thus offenses were violations of religious law rather than municipal ordinance.

In Chinese cities, by 900 A.D., the guilds provided street cleaning services (garbage collection) for their own streets.[9] By 1250 A.D., in the city of Hangchow, the streets were cleansed by public authorities who had the refuse removed to the countryside by boat, and once a year the prefecture undertook a thorough cleansing of the streets and a general cleaning out of the canals.[10]

The Medieval Era

By the thirteenth century the larger European cities were again coming to terms with the solid waste problem after several centuries of neglect. In their case two factors aggravated the problem: while the population was steadily increasing, a generally constant city size (area) was dictated by defensive walls, and the population tended to maintain rural modes of life, including keeping animals, under these crowded conditions. On the other hand, medieval refuse was mostly organic, save such things as tannery wastes and ashes, and tended eventually to decompose and mix with the earth. As in earlier times, citizens had the duty of scavenging, though this was often ignored and most refuse was simply thrown into the streets. Dogs and pigs were, in general, the only effective scavengers (as Mumford notes,[11] a 1217 miniature shows a sheep and a pig crossing a bridge in Paris, then the largest city in Europe).

However, some scavenging must have been done, for by the year 1240 small mountains of refuse had been formed just outside the gates of Paris. These would become a part of Parisian topography. (By 1400 these dumps towered over the city walls and were rightly seen as a weakness in the city's defenses. When the Portuguese were attacking the Morrocan town of Ceuta in 1415, two Portuguese princes gallantly charged up a dominating mound, which seemed to be a strategic point. They soon discovered that they had thus heroically captured the city dump.)

As early as 1243, the magistrates of Avignon decreed that no one should throw any refuse into the street, on the pain of a fine of two shillings per offense, of which the accuser would receive one third. Aldermen in London were each empowered to appoint "four reputable men" to clean and repair the street (1280). On the other hand, Siena in 1296 officially employed a sow with four young pigs to clean the campo after every market; by 1382 six pigs had the official duty of cleaning the streets. Venice and Florence also raised pigs especially for scavenging, while in Bologna the Ospitalieri di Sant' Antonio had the privilege of feeding one hundred pigs in the streets.

St. Omer in France is recorded as having taken special care to keep the streets and square as clean as possible. They propagated an ingenious regulation according to which whoever brought a cart of sand, earth, gravel or sod into the city should, in exchange, carry out of the city a load of mud or ordure.

The primary measures taken in other contemporary large cities were street paving (Paris, c. 1320), canalization of wastes (open sewers) and, of course, organized scavenging.

Toward the end of the Middle Ages similar measures and a general awareness of solid waste as a *health* problem (as opposed to a *transportation* problem) became quite common. While Edward the Third of England's Proclamation of 1372 that "throwing of rushes, dung, refuse and other filth and harmful things into the [Thames] shall no longer be allowed" was to ease the passage of ships, the act of Parliament in 1388 which "forbade the throwing of filth and garbage into ditches, rivers, and waters" was the result of over a century of lobbying by those—noble and commoner—who lived in the more afflicted areas.[12] (The latter decree was in part a response to the butchers' practice of dragging their filth through the streets and dumping it into the Thames.)[13]

Furthermore, by this time London had official scavengers controlled by a surveyor of the streets (although there were only twelve scavengers' carts for all of London—which carried garbage either to dumps outside the walls or to barges to be ferried away). Other European cities adopted similar measures. Milan had a board which specifically dealt with street cleaning and environmental sanitation and in several cities in southern Germany (for example, Nuremberg) scavenging and street cleaning became public services. In 1462 the aldermen in Amiens decided to hire certain men with horses and dung carts to remove the filth from the streets daily.

Ultimately, legislation and scavenging tended to be relatively ineffective. This was not, however, because of ignorance but rather because offenders and offended alike were unable to devise adequate alternatives to the available methods of collection and disposal. Moreover, except for those living in heavily polluted areas, popular opinion was very much against such measures. ("Should the poor man be deprived of his pigs to make way for the horses of the wealthy? Should the poor man be forced to spend his time carting garbage in order that lords and ladies not soil their costly garments?")[14]

The Fifteenth to Eighteenth Centuries

Any progress during the next few centuries was quite gradual. (It was not until the nineteenth century that what we now regard as acceptable methods of dealing with the solid waste problem began to be developed.) Even the rudimentary measures noted were found only in major cities. In the cities of the provinces, street paving was confined to the main streets, and while rich citizens might possess a courtyard into which all rubbish was thrown, and occasionally removed to the suburbs, ordinary citizens still threw everything into the streets. In Naples the breakwater which protected moored ships became so encumbered by refuse that in 1597 serious considerations had to be given not to cleaning it, but to replacing it with a new breakwater.

By and large, the measures taken did improve, partially in response to the series of plagues which struck Europe between 1350 and 1750. In England, Henry VI established a Commission of Sewers during the fifteenth century, which in turn provided for severe penalties for the pollution of streams and made special provisions for the disposal of tanners' and brewers' wastes. Henry VII forbade slaughterhouses in cities or towns "leste it might engendre sickness, unto the destruction of the people."[15] Citizens throwing rubbish into public places were made liable to a fine of forty shillings.

Rouen during the sixteenth century hired carters to clean the streets and transport their sweepings to dumping grounds under the walls. The inhabitants were obliged to clean the gutter in the middle of the road and the streets in front of their houses. In Oxford, citizens who allowed swine to roam in the streets were fined (1532) and a "Comon Skavinger" was appointed (1541). It is reputed that Shakespeare's father in 1552 paid a fine for depositing filth in a public street. It is also said that the Spanish city of Valencia purposefully did not pave its streets "because their refuse mixed with the excrement with which [the streets] are only strewn for a few moments is carried at frequent intervals to fertilize the adjoining countryside"[16] and the people [were] convinced that if they were to pave them, they would deprive "the great orchard, which surrounds Valencia on all sides, of one of the principal sources of its fertility."[17]

After the great fire of London in 1966, laystalls, or temporary dumping places, were set up in the streets and orders were given that all wastes should be deposited there for later removal. Officials were appointed to supervise the cleaning of the streets. (These officials were called "scavengers," but the position was both unpaid and supervisory.) The great part of the work was done by rakers. However, "as the City streets became increasingly built up, sites for laystalls became more difficult to find, and they were later replaced by fixed or moveable boxes into which refuse was thrown."[18] A charge was imposed on property owners for removal of rubbish by the rakers, who later "became paid servants of the commissioners of sewers and afterwards of the City Corporation."[19]

Paris in 1666 was not quite as exemplary. That year, as the area under the Pont Marie was a favorite dumping ground, the city magistrate had to order the residents to remove the rubbish to a dump because the Seine was no longer able to move freely under the arches and was forced against, and thus endangered, the bridge itself. "Typical was the magistrate's reminder that ordinances forbade garbage disposal on quays or river banks."[20] Nevertheless, when, late in the sixteenth century, Henry the Fourth tried to reorganize municipal disposal of wastes, his efforts foundered on public indifference. Furthermore, even tanners and dyers were not ousted from the city until 1673. Tokyo, on the other hand, systematized garbage collection, licensed refuse contractors to collect and dispose of home wastes, and provided a dumping ground as early as 1662. In the Netherlands, dumping garbage in the canals was first prohibited by guild bylaws, so that the breweries should have a clean supply of water.

Several provincial English towns provided collection and fenced off dumps by the end of the seventeenth century. In 1690, the Mayor of Portsmouth was fined for throwing garbage into the streets of his own town. An English law of 1714 required every town to engage a scavenger.

The European settlement of North America brought with it the solid waste problems, and methods of dealing with these problems, which were prevalent in Europe at that time. Initially, the low population density of these settlers enabled them to get rid of refuse however they wished without any repercussions, and this continued to apply on what became known as the frontier. (The native population in what is now the United States had not built any cities with populations much over 10,000. Mesoamerican civilizations, on the other hand, had maintained complex urban systems—including sophisticated drainage and waste collection—for at least a thousand years.)

Urban concentrations soon sprang up, both for reasons of defense and maintenance of commerce. Local solutions tended naturally to reflect the national background of that region's settlers. For example, the Dutch settlers in what is now Manhattan were exemplary in their treatment of solid waste, much as were their ancestors.

As early as 1644 (eighteen years after the Dutch bought Manhattan Island from the Indians) the residents were directed to take all wastes out of the fort; and in 1648 a law was passed prohibiting hogs and goats from running in the streets. (This was not successful, because these animals were not only effective as scavengers, but also were a cheap source of food for the poor. Indeed, though many similar ordinances were passed in ensuing years, it was not until the mid-nineteenth century that animals other than horses were effectively banned from the streets of New York City.) An ordinance in 1657 prohibited the throwing of wastes into the streets and specified four sites to which these had to be taken; in addition, the inhabitants were made responsible for the condition of the streets in front of their lots.

A collection system was established in 1670 (following the orderly transition of power from the Dutch to the British in 1664) when cartmen, in exchange for a monopoly, agreed to clean all rubbish from the streets once a week. Prior to 1691, the citizen had to load his own wastes onto the cart, but thereafter he could pay extra to have this done for him. However, in 1676, New York residents had a foretaste of what was to become a bane of urban existence, municipal strikes, when the cartmen refused to move garbage from the streets to protest what they considered too low a rate of pay.

This arrangement persisted throughout most of the eighteenth century and was so effective that there was universal agreement among travel writers of the day that New York, as compared with other cities both in Europe and in America, was very clean. Boston, however, was not highly regarded. Although ordinances of 1634 and 1652 forbade throwing any wastes onto "any hie way or dich or Common," it was not until 1667-68 that the annual town meeting, perhaps prompted by the current smallpox epidemic, for the first time elected scavengers. As in London, this job was unpaid and supervisory. One reason for this attitude was doubtless the strict biblical interpretations favored by the Puritans. Blake says, "The fact that seventeenth century therapeutics so often proved unavailing was undoubtedly one of the circumstances conditioning this reaction to sickness [regarding it as the 'scourge of God']. On the other hand, it would also appear that this attitude in turn inhibited aggressive application of existing knowledge and beliefs for the protection of the public health."[21] Not until 1712-13 did ordinances empower the scavengers to hire men and carts to clean the streets as often as they thought necessary, but because of a taxpayer's protest only carts were provided the next year, and nothing at all the year after that. Five years later this arrangement was again tried, but it was not until 1797, after the War of Independence, that municipal collection of garbage was begun for good, when one man with a cart was hired by the selectmen to carry away street dirt. Two years later, in response to a yellow fever epidemic, the newly founded Board of Health forbade the throwing of refuse into waterways or onto docks and provided "six horse drawn carts to go through each ward twice a week [soon three times] from May 1 to November 1 to remove all the dirt ... collected by the inhabitants."[22] Despite complaints, the board was not willing to increase the collection force to more than nine carts. "Moreover, the board was never able to devise a satisfactory method of disposing of the refuse after it was collected, though many expedients were tried, including an effort to provide sanitary fill at the then marshy border of the Common."[23]

In Charles Town, North Carolina, an ordinance providing for a scavenger in charge of street cleaning was passed in 1744, and a system of collection was provided for in 1750. Five years later it became illegal, under threat of a fine, to dirty the streets after they had been cleaned. The Kennebunkport, Maine, municipal dump dates from 1776.

However, few other American cities had either organization or facilities for refuse collection and disposal prior to the 1790s. Not until 1792 did Philadelphia initiate its system, whereby slaves would carry loads of garbage on their heads, to be dumped in the Delaware River downstream from the city (according to a plan of Benjamin Franklin's).

Europe during the eighteenth century showed little of the change that was to characterize the next century. In Edinburgh, regaled as "the filthiest city in all Europe,"[24] all garbage was thrown into the street at night and was not removed until early in the morning when it was perfunctorily cleared away by the city guard. On Sabbath morn, however, it could not be touched, and lay there all day long, filling Scotland's capital with the "savour of mistaken piety."[25] (Recent excavations have shown that "modern heavy traffic [in Edinburgh] is carried on a thickness of consolidated refuse which sometimes reaches a depth of fifteen feet.")[26] Nevertheless, in provincial England, perhaps influenced by mercantilism (national self-aggrandizement requires a sound labor force, and thus some sort of public health policy), 211 acts for paving and other parochial improvements were passed between 1785 and 1800. These gave to vestries or special bodies of commissioners powers in regard to paving, cleaning, or lighting; some required the householder to clean the pavement in front of his property, and others enabled authorities to make arrangements with rakers for streetcleaning and the removal of dirt and rubbish. "In those days, and even in more recent times, scavenging contractors were willing to pay for the privilege of sweeping the streets and collecting the sweepings and household refuse . . . ," which they then sold to nearby farmers.[27] In Bristol, by 1794, the streets were cleaned twice a week, footways every morning, and there was a full-time Inspector of Nuisances.

The Lord Mayor of Dublin in 1805 issued a proclamation instructing that, in order to protect the people from the diseases incidental to filth, he would pay them to remove it. In eighteenth-century Paris, on the other hand, there were riots because the accumulation of filth on the streets was so great.

The Nineteenth and Twentieth Centuries

The nineteenth century marks the beginning, at least in general terms, of all modern concepts of solid waste collection and disposal. While the century is commonly associated with laissez-faire capitalism, it was equally, as Beatrice and Stanley Webb pointed out, the century of municipal socialism. "Neither a pure water supply nor the collective disposal of garbage . . . could be left to private conscience or attended to only if they could be provided for at a profit."[28] As opposed to the perceived need for a healthy labor force which dominated eighteenth-century public health thinking, the nineteenth century reformers were moved by the philosophy of Bentham and the Utilitarians, according to whom the supreme ethical and political principle was the maximization of the "happiness" of the population as a whole.

Nevertheless, conditions in working class towns reached "a pitch of foulness and filth . . . that the lowest serf's cottage scarcely achieved in medieval Europe."[29] Workers' houses would be built right up against a factory or railroad. While for several centuries it had been an offense to throw rubbish into the streets in English towns, in these industrial cities this was the regular means of disposal. "The rubbish remained there, no matter how vile and filthy, until the accumulation induced someone to carry it away for manure."[30] As late as 1840, some of the poor in Glasgow stored their wastes in outdoor piles for sale as fertilizer as a means of defraying their rent. While most of the other towns in Europe, and some in the United States, were developing some sort of universal solid waste collection system, a royal commission in England in 1844 found only two towns where the collection system extended into the slums.

New York City was one of the few nonindustrial cities where conditions became worse instead of better in the nineteenth century. Problems arising from years of neglect during the disruption of urban government in the Revolutionary period were compounded by an explosive population increase. Despite repeated attempts to legislate improvement, "almost every aspect of sanitary conditions was worse in 1825 than it had been at the end of the century . . . Instead of developing an effective garbage collection system, the City alternated between private contractors and a city-operated program, neither of which worked satisfactorily. A major part of the scavenging came to be done by the ever-present hogs."[31]

After the Civil War, inflation compounded the difficulty of cleaning up the cities. Contractors broke contracts as currency became less valuable. Conditions were worst in the bankrupt southern towns such as Memphis and New Orleans, where facilities were primitive and municipal organization sometimes nonexistent. In New York, however, the Board of Health, apprehensive over a possible outbreak of Asiatic cholera, tried to bring order to the chaotic sanitary system. Two hundred and nine piggeries were evicted from the city and 3,949 refuse-heaped backyards were cleaned up. The first watertight garbage cans were introduced, and collectors were prevailed upon to make regular rounds with sturdier vehicles. Nevertheless, "Boss" Tweed had let out all contracts on a ten-year basis, and by 1871 the situation had again grown so bad that many streets were privately cleaned through voluntary contributions.

The New York City experience in the nineteenth century, while severe, is still indicative of the general experience of most major urban areas. Duffy says:

> From the eighteenth century to the present influx of southern Blacks and Puerto Ricans, the [economic and cultural backgrounds of most immigrants arriving in New York have been relatively low]. Moreover, since a democratic government reflects the quality of the electorate, the city's sanitary problems were compounded by a relatively inefficient municipal system. Street cleaning and garbage removal in nineteenth century American cities involved far more than a simple business or economic operation to remove wastes and debris. The contracts provided jobs for constituents, cash for political slush funds, a supplementary source of income for politicians, and political favors for their friends.[32]

Thus while the political reform movements of the latter part of the century did not perceive refuse collection and disposal as an area of concern, their efforts had a definite, positive influence on the quality of such service.

In England, the Sanitary Act of 1845, and especially the Public Health Act of 1875, had a radical effect on the general level of sanitation, notably in working-class urban areas; they included provisions for the creation of "united districts" for "specified sanitary purposes." Tokyo, after the Meiji Succession, required each family to furnish a garbage box in 1807, and began municipal collection in 1900. In New York City, Colonel George E. Waring was appointed commissioner of the Department of Street Cleaning in 1894 and, untroubled by political considerations, proceeded to reorganize and clean up the Department. Still, by the late nineteenth century, in St. Louis and Baltimore, much of the refuse was simply carted to open dumps, and Cleveland had no provision for gathering household garbage at all.

The latter part of the nineteenth century also marked the beginning of the technological approach to solid waste collection and disposal. Prior to that time disposal especially posed a difficult problem. In many towns the contents of privies and domestic ashpits were collected at the same time as the street sweepings, and the absolute amount of wastes was rapidly increasing because of the increase in urban population and changes in consumption. Liverpool in 1866 produced about 2000 tons of refuse daily; Manhattan in 1900 produced about 600. Some refuse was sold to farmers as fertilizer, and sometimes markets could be found for other wastes. Until 1884, Boston sold all its garbage to New England farmers. St. Paul, Denver, and Omaha sold garbage to local farmers for a longer time, and Philadelphia still did in the 1970s. Worcester, Massachusetts, maintained its own hog farm, which, in 1903, paid two-thirds of the cost of municipal collection. (This practice was discontinued for health reasons—for instance, in 1889 thirteen percent of the hogs fed on Boston garbage were found to have trichinosis.) Nevertheless, there still remained a large quantity of garbage to be disposed of, and this was usually done by open dumping into the sea or onto wasteland.

Although garbage had been burned for some time, the invention of the incinerator (in England, in 1874) was a marked step forward in the clean disposal of solid wastes. While the disposal problem was not eliminated, it was definitely reduced. The first municipal incinerator in the United States was built in Allegheny, Pennsylvania, and by 1902 an MIT survey of solid waste practices[33] could report that, of the 183 largest cities, 15 percent used some system of incineration. However, only 10.4 percent used some other method of reduction or utilization, and only 4.9 percent dumped in dumps, so that fully 69 percent relied on more primitive methods of disposal, for example, open dumping on land or into water, or selling garbage as hog feed. (On the other hand, 64 percent of the cities surveyed had either municipal or contracted private collection, and only 7.4 percent had no systematic collection of wastes.)

The search for technological solutions became relatively more intensive in the twentieth century. Garbage grinding was invented in the '20s, compactor trucks in the '30s. From 1913 to 1959, Chicago had a privately owned tunnel system which removed building debris and ashes as well as delivered freight. More recently, containerized pickup has become common, and advanced systems such as the vacuum collection system on Roosevelt Island in New York are being introduced.

Advanced management methods are also being introduced into this service industry. Improved procedures have been employed to schedule work crews,[34] route refuse collection vehicles,[35] and to locate disposal facilities.[36] Major technological advances are taking place in the recovery and recycling of materials,[37] and in the use of solid waste as a fuel.[38]

But this evolution in the handling of solid waste has occurred primarily in the industrialized nations. The developing countries, especially those with large urban areas, have extreme problems in dealing with their solid wastes; Calcutta is an example: "Garbage piles so high in the streets that the Government regularly recruits emergency volunteer forces to reduce it so that people can move around." Nineteenth-century refuse practices are utterly unable to cope with twentieth-century populations—in Calcutta, eight million people, 177,000 per square mile. "The basic difficulty is that the great bulk of [the] rush to the cities is concentrated in those nations of Asia, Africa, and Latin America which are poorest in the financial, human, and particularly administrative resources needed to cope."[39]

It would be a mistake to assume too much proficiency on our own part. Los Angeles collected garbage by mule team until 1933, and, as recently as 1941, several New Yorkers were prosecuted for keeping hogs.

Conclusion

Man has struggled with the problem of solid waste management for six thousand years. The recurring issues are how to remove the wastes from his midst, how to assure compliance on the part of all citizens, who should perform these duties, and how to pay for them. The next chapter provides a framework for addressing these issues in a modern city.

Notes

1. Lewis Mumford, *The City in History: Its Origins, Its Transformation, and Its Prospects* (New York: Harcourt, Brace & World, 1961).

2. Charles Joseph Singer et al., eds., *A History of Technology*, 5 vols. (Oxford: Clarendon Press, 1954-1958), vols. 1 and 4.

3. Michael A. Hoffmann, "The Social Context of Trash Disposal in an Early Dynastic Egyptian Town," *American Antiquity* 39, no. 1 (January 1974): 35-50.

4. Deut. 23: 13-15.

5. *Troy; excavations conducted by the University of Cincinnati, 1932-1938*, ed. Carl William Blegen, 4 vols. (Princeton: Princeton University Press, 1950-1958), vol. 1.

6. Arnold Hugh Martin Jones, *The Greek City from Alexander to Justinian* (Oxford: Clarendon Press, 1940).

7. Benjamin R. Foster, "Agoranomos and Muhtasib," *Journal of the Economic and Social History of the Orient*, vol. 13, pt. 2 (April 1970): 128-144.

8. G.T. Scanlon, "Housing and Sanitation: Some Aspects of Medieval Islamic Public Service," in *The Islamic City: a Colloquium*, ed. Albert Habib Hourani and S.M. Stern (Philadelphia: University of Pennsylvania Press, 1970).

9. Wolfram Eberhard, *A History of China*, trans. E.W. Dickes (Berkeley: University of California Press, 1960).

10. Jacques Gernet, *Daily Life in China, on the Eve of the Mongol Invasion, 1250-1276*, trans. H.M. Wright (Stanford: Stanford University Press, 1962).

11. Mumford, *City in History*.

12. Ibid.

13. Ernest L. Sabine, "Butchering in Medieval London," *Speculum; a Journal of Medieval Studies* 8 (1933): 335-353.

14. Mabel Craven Buer, *Health, Wealth, and Population in the Early Days of the Industrial Revolution* (New York: H. Fertig, 1968).

15. John Ballard Blake, *Public Health in the Town of Boston, 1630-1822*, Harvard Historical Studies, vol. 72 (Cambridge: Harvard University Press, 1959).

16. Sabine, "Butchering in Medieval London."

17. Fernand Braudel, *The Mediterranean and the Mediterranean World in the Age of Phillip II*, trans. Sian Reynolds, 2 vols. (New York: Harper & Row, 1975).

18. J. Rawlinson, "Sanitary Engineering," in Singer, *History of Technology*, vol. 3.

19. Ibid.

20. Andrew P. Trout, "The Quai Pelletier," *History Today* 23 (December 1973): 858-863.

21. Blake, *Public Health in Boston*.

22. Ibid.

23. Ibid.

24. George Macaulay Trevelyan, *English Social History; a Survey of Six Centuries, Chaucer to Queen Victoria* (London, New York: Longmans, Green, 1944).

25. Ibid.

26. John Capie Wylie, *The Wastes of Civilization* (London: Faber and Faber, 1959).

27. Rawlinson, "Sanitary Engineering."

28. Mumford, *City in History*.

29. Ibid.

30. Ibid.

31. John Duffy, *A History of Public Health in New York City*, 2 vols. (New York: Russell Sage Foundation, 1968-1974).

32. Ibid.

33. William Francis Morse, "The Collection and Disposal of Municipal Waste," *Municipal Journal and Engineer*, New York, 1908.

34. Stanley Altman et al., "A Nonlinear Programming Model of Crew Assignments for Household Refuse Collection," *IEEE Transactions on Systems, Man and Cybernetics*, SMC-1, no. 1 (July 1971).

35. Richard M. Bodner, E. Alan Cassell, and Peter J. Andros, "Optimal Routing of Refuse Collection Vehicles," *Journal of the Sanitary Engineering Division, American Society of Civil Engineers* 96 (1970): 893-904.

36. Peter Thursfield and J. Fairley, *Refuse Disposal in North East Cheshire* (Reading, Eng.: Local Government Operational Research Unit, Royal Institute of Public Administration, 1969).

37. Helmut Schulz et al., *A Pollution-Free System for the Economic Utilization of Municipal Solid Waste* (New York: School of Engineering and Applied Sciences, Columbia University, 1973).

38. Richard C. Bailie, "Solid Waste as an Energy Resource," *Professional Engineer* 42 (February 1972): 42-47.

39. *New York Times*, 9 June 1975, p. 36.

3 The Organization of Solid Waste Collection: A Framework for Analysis

Municipal solid waste can be removed from the premises where it is generated in several ways: (1) It can be ground up by garbage disposal units ("garbage grinders") into a slurry and discharged into the wastewater (sewer) system. (2) It can be conveyed pneumatically through tubes to a central collection point; (this is a recent technological development, particularly applicable to densely situated multiple dwellings). (3) It can be placed in suitable containers which are emptied into or carried away by an appropriate vehicle. (Street sweeping fits into this category, if one considers the hopper of a mechanical sweeper as the container in question.) The first two methods are insignificant in terms of the total amount of solid waste removed; the third method is the only one of any quantitative significance today and in the foreseeable future in metropolitan areas. Such collection is the most important means of removing municipal solid waste, and this class of waste constitutes the majority by weight of all solid waste that is formally collected. The subsequent discussion therefore is addressed to this method of collection, and the first requirement is to define the various organizational arrangements that can be employed in this activity.

Service Elements

The organizational structure for delivery of services can be specified by four elements or parameters: the service recipient, the service provider, the service arranger, and the service type. (The method of payment for service could be considered a fifth parameter, but instead it will be treated as a separate topic in a later chapter.) Each of the four elements is discussed in turn.

Service Recipient

The recipient of solid waste collection service is either (1) a residential household, which may be in either a large multiple dwelling or a small structure of one to several housing units—many cities divide their residential structures into these two classes and apply different collection arrangements to them; (2) a commercial establishment, such as a restaurant, office building, store, or supermarket; (3) an institution, such as a school, hospital, jail, or nursing home; (4) an industrial establishment, such as a steel mill or a factory; or, (5) the streets, parks, and other public spaces in a community, which can be said to "produce" the component of municipal solid waste identified as litter and street sweepings.

Service Provider

The service provider physically collects solid waste from the service recipient, that is, removes it from his premises. For municipal solid waste, the following six important service providers can be identified:

1. A local government bureau, often within a sanitation department or a department of public works, of the jurisdiction (municipality or township) in which the service recipient is located;
2. A unit of general-purpose government other than the municipality or township in which the service recipient is located, such as a county government unit or even a unit of *another* municipality;
3. A special district or authority created by state or local law;
4. The service recipient himself; for example, in many communities the individual householder carries his garbage and trash to the local disposal facility. Similarly, a supermarket chain might assign a truck and two employees to service the units of the chain that are located within a reasonable area;
5. A private firm engaged in the business of collecting solid waste. Such firms are identified in various ways in various places. Most often they are called garbage, refuse, rubbish, trash, or waste removal firms, or garbage, refuse, rubbish, trash, or waste haulers or contractors. Sometimes they are called carters or scavengers. The firm may simply provide service to individual recipients who wish to purchase it, it may be under contract to a locality, or it may have a franchise awarded by the locality. The latter will be referred to in subsequent discussion as a franchise or a franchised firm. The distinguishing characteristic of such a firm is that service recipients (of a given class) located within the franchise territory act as service arrangers but are not permitted to select any other organization to provide service delivery. Indeed, if the locality has a mandatory collection law, which requires each waste generator (service recipient) to purchase regular service from a service provider, the service recipient may not even act as his own service provider; he must arrange for service from the franchisee for this work.
6. A voluntary association; for instance, a neighborhood recycling center or a Boy Scout troop, while not a unit of government, might collect newspapers and aluminum cans.

Service Arranger

The concept of service arranger is less obvious than the concepts of service recipient and service provider and requires careful explication. The service arranger is the agent who assigns the service provider to the service recipient. The five important service arrangers for the collection of municipal solid waste are:

1. The municipality or township or county where the service recipient is located; the municipality may decide to use a government agency or a private firm (as service provider) to collect its residents' solid waste.
2. The county where the service recipient is located; the county has the same options described above for the municipality.
3. The service recipient himself, as in the case of a resident or commercial establishment that decides to hire a private firm to haul away its trash.
4. A special district or authority, which, like a municipality, may choose to use its own work force or a private firm as the service provider for the service recipients within its jurisdiction.
5. A voluntary association, such as a neighborhood civic group, may arrange for its own collecting or may contract with a private hauling firm to collect solid waste from the homes or businesses of members of the association. The latter is not uncommon in unincorporated areas.

Service Type

The final parameter needed to define the organization pattern is the type of service. This term refers to the type of solid waste collected. Garbage (putrescible waste), dry trash, yard trash, bulk waste, leaves, newspapers, and mixed residential refuse are examples of types of residential solid waste. For commercial establishments and institutions it is useful to distinguish among certain types of service: for garbage, corrugated paperboard, combustible trash, and so on.

Different service providers and arrangers may be used for the different types of service. It would not be surprising, for example, to have the municipality as service arranger and provider for residential garbage and dry trash collection in a community, while the service recipient acts as service arranger and hires a private firm as service provider for the collection of his bulk waste, and at the same time acts as his own service provider and brings his yard trash to a composting site. Further, the homeowner may authorize the local Boy Scout troop, a voluntary association, to collect his old newspapers for recycling.

Table 3-1 summarizes the different elements of this organizational framework. It is clear that a four-dimensional matrix, one dimension for each of the

Table 3-1
Organizational Elements in Solid Waste Collection

Service Recipient	residential household
	commercial establishment
	industrial establishment
	institution (schools, hospitals, religious structures)
	streets (roadways, sidewalks, litter baskets)
	parks and other public spaces
Service Provider	municipal government
	county government
	local government of a different jurisdiction
	private hauling firm
	special district or authority
	service recipient
	voluntary association
Service Arranger	municipality
	county
	special district or authority
	voluntary association
	service recipient
Type of Service	collection of mixed residential refuse
	collection of garbage
	collection of trash
	collection of yard trash
	collection of combustibles
	collection of noncombustibles
	collection of bulk
	collection of paper

four elements discussed, is needed to describe completely a community's organizational pattern for collecting municipal solid waste. It is therefore obvious that the situation is considerably more complex than the simplistic, one-dimensional classification of communities as municipal, contract, private (and combinations of these) that have been conventionally used heretofore. For example, a city described by the American Public Works Association[1] as having "municipal and private collection" could be one where service is provided to all one-family dwellings by a municipal department, while all other collection is carried out for a fee by a single private firm which has an exclusive long-term franchise. On the other hand, the same inadequate term, *municipal and private,* would describe a city where a municipal department collects only residential garbage once a week in one area of the city while several private firms compete vigorously and offer a wide array of services to all households and commercial establishments.

Service Arrangements

The elements of solid waste collection service having been enumerated, the next step is to identify service arrangements. A means of simplifying the large number of combinations that could, in theory, result from the list of elements is to pair service providers and arrangers, to call each useful pair an "arrangement," and then to examine the different arrangements used for various recipients and types of service.

Table 3-1 shows seven kinds of service providers and five kinds of service arrangers. In principal, therefore, thirty-five different pairings could exist. However, many of the conceivable pairings are impractical or meaningless. For example, neither government nor a special authority nor a voluntary association can demand that a service recipient haul his own refuse, although they can demand that he arrange for its removal; if he chooses to haul his own, the arrangement is properly termed self-service in this scheme. Furthermore, it is difficult—although not impossible—to conceive of the circumstances in which a special district would hire a local government unit to provide service to member communities; a direct relationship between communities via an intergovernmental contract arrangement is more likely.

Service arrangements are shown in Table 3-2 and twelve arrangement types are named there. Specific cities which utilize the indicated arrangement, at least to some extent, for the collection of residential mixed refuse are also identified for illustrative purposes. A more complete list of arrangements is defined in Table 3-3. The names in the table will be used throughout this volume as shorthand for the indicated pairs of service arrangers and providers. The terms *municipal arrangement, municipal collection,* and *municipal service* will be employed interchangeably, as will those for the other arrangements.

There may be a certain amount of ambiguity as to the identity of the service arranger when the service provider is a private firm, particularly in the case of residential collection, inasmuch as both the municipality and the service recipient (household) may play a role. Three common arrangements—municipal, contract, and franchise—can be distinguished by reference to two conditions: the presence or absence of a contractual relationship between the municipality and the private firm and whether the municipality or the firm collects money from the service recipient for the service. Table 3-4 defines the various conditions and the resulting arrangements.

As will be seen in the next chapter, municipal, contract, franchise, private, and self-service arrangements account for the overwhelming majority of arrange-

Table 3-2
Service Arrangements for Solid Waste Collection

	Service Provider					
Service Arranger	Local Municipality	County	Another Government	Special District	Private Firm	Service Recipient	Voluntary Association
Municipality	municipal (New York)	intergovernmental contract	intergovernmental contract	(Des Moines)	contract (Boston) franchise—mandatory (San Francisco)	NA	
County		county				NA	
Special District	NA	NA	NA	special district (Indianapolis)	special district contract (Jersey City)	NA	
Voluntary Association					association contract (Houston)	NA	voluntary association
Service Recipient					franchise—nonmandatory private (St. Paul)	self-service (Eugene, Ore.)	voluntary service

Table 3-3
Nomenclature and Definition of Service Arrangements

Service Arranger	Service Provider	Name of Arrangement
municipality (or township)	municipality (or township)	municipal
municipality (or township)	private firm	contract
service recipient or municipality	private firm with exclusive franchise	franchise
service recipient	private firm without exclusive franchise	private
service recipient	service recipient	self-service
service recipient	voluntary association	voluntary service
county	county	county
municipality (or township)	another government	intergovernmental contract
special district (or authority)	special district (or authority)	special district
special district (or authority)	private firm	special district—contract
municipality (or township)	special district (or authority)	municipality—special district
voluntary association	voluntary association	voluntary association
voluntary association	private firm	association contract

ments in effect for residential, commercial, and institutional solid waste collection.

Multiple Service Arrangements

Multiple arrangements can and do exist within a community. Even if attention is restricted to a single class of service recipient, the household, and a single service type, namely, mixed residential refuse collection, several different arrangements or arranger-provider pairings can coexist. In Indianapolis, for example, four

Table 3-4
Definition of Contract, Franchise, and Private Arrangements Where the Service Provider Is a Private Firm

		Who collects payment from service recipient?	
		Firm	*Municipality*
Is there a contractual relationship between the municipality and the private firm?	Yes	franchise	contract
	No	private	not applicable

arrangements can be found, as shown by the four entries in the main part of Table 3-5 (which is an abbreviated version of Table 3-2: because our attention at this point is focused on only two of the four dimensions, a two-dimensional table suffices). A special authority, the Indianapolis Sanitary District (ISD), has its own work force and collects waste from most households, but it also contracts with private firms to provide service to households in certain areas of the city. Nevertheless, there are many households that lie outside the legislatively prescribed ISD territory but inside the city limits. Those households are free to hire private firms or to deliver their own refuse to a disposal facility, and both of these arrangements are found.

The entries in the table refer to the estimated fraction of households serviced by each arrangement. This kind of quantitative information is usually difficult to obtain. Although many cities have multiple arrangements, there is often considerable uncertainty in such cities about the number of recipients serviced by each arrangement; generally speaking, public officials do not have complete information about this.

Multiple arrangements can also be found with respect to commercial establishments. For example, in some cities a municipal agency collects from stores and offices on city streets, but not from similar establishments located in shopping centers; the latter must make other arrangements, either for service by a private hauler or for self-service. A similar situation prevails in a community where the municipality collects from stores, but only twice a week, for example. This may be sufficient for some establishments (for example, dry cleaning) but not for others (as restaurants). The latter would then make other arrangements for the higher frequency of service they require.

The presence of multiple arrangements for the same class of service recipient and the same type of waste, as in Indianapolis, in no way should be interpreted as poor organization. Indeed, it is not unreasonable to speculate that the existence of alternative arrangements in the same city can lead to healthy competition and provides standards for comparisons that are lacking in cities that limit themselves to only a single arrangement.

Table 3-5
Municipal Service Arrangements in Indianapolis for Collection of Mixed Residential Refuse

	Service Provider			
Service Arranger	ISD	Private Firms	Service Recipient	Totals
Indianapolis Sanitary District	70%	7%		77%
Service Recipient		18%	5%	23%
Totals	70%	25%	5%	100%

Service Patterns

At this point, having defined the elements of service structures, having defined an arrangement as a service arranger-provider pair, and having noted the use of multiple arrangements, one can synthesize a framework that incorporates all these concepts. For any city, the entire *pattern* of service arrangements, service recipients, and service types, can be displayed in the form of a matrix, as shown in Table 3-6. (Minor variations in this matrix would allow for the different types of service—glass, aluminum cans, and so on—which may apply to different communities.) The use of such a matrix facilitates comparisons between cities, helps identify the broad range of options available to a community, and affords a full but clear portrayal of a city's whole complex pattern of collection services for municipal solid waste. To illustrate the use of the table, the service pattern for the city of Indianapolis in 1974 is portrayed by the entries in the exhibit.

Summary

A framework is developed for describing the institutional structure of solid waste collection services in a community. It requires identification of the kind of organization that *arranges* for the service and the kind of organization that *provides* the service to each class of *service recipient,* for each *type of service.* To this information must be added information on the *fraction* (or number) of service recipients who are served by each arrangement.

Specifically, with regard to solid waste collection, a number of important organization *arrangements,* that is, pairs of service arrangers and service providers, are identified. They can be utilized, singly or in combination, to supply the various types of service to the different classes of service recipients, thereby creating a *pattern* of arrangements for solid waste collection services within a jurisdiction.

Note

1. American Public Works Association, *Refuse Collection Practice*, 3rd ed. (Chicago: Public Administration Service, 1966).

Table 3-6
Pattern of Municipal Solid Waste Collection Services for a City

Service Arrangement	Service Recipient								
	Multiple Dwellings	Other Residential Dwellings						Commercial	Institutional	Streets and Public Property
		Mixed Refuse	Type of Service							
			Wet Garbage	Yard Trash	Separated					
					Bulk Waste	Newspapers	Other			
Municipal										100%
Contract										
Franchise										
Private	X	18%			X			X		
Self-Service		5%			X			X	X	
Special District	X	70%			X				X	
Special District–Contract		7%								
Intergovernmental Contract										
Voluntary Service						X				
Other										

4

The Organization of Solid Waste Collection: Findings

Using the framework developed in the preceding chapter, this chapter presents findings on the extent to which the different arrangements are employed. First, it reviews important previous studies on this subject. Then it describes the sampling procedure used to determine the current organization patterns. Next, the principal findings are presented: the various service recipients and their respective service providers, service arrangements, types of service, and reported changes in service arrangements. Finally, the findings are extrapolated and summarized.

Previous Studies

There were eleven major studies of organizational arrangements for solid waste collection prior to the study reported here. (See Table 4-1.) It is difficult to generalize from them, however, because of differences in sampling technique, in definitions, and even in service recipients and service types included in the studies. Nonetheless, it is useful to review those studies; a tabular summary appears at the end of this section.

1902 MIT Survey

The earliest of these surveys was conducted by MIT in 1902, and was reported by Morse.[1] It covered 161 "representative" cities in the United States, with populations greater than 28,000. Of the 155 cities reported, 29 were said to have no systematic method of collection. However, 146 cities reported a collection method; the apparent discrepancy may be explained by the fact that several cities had a nonsystematic arrangement, that is, in response to demand. Of these 146 cities, 61 (or 42 percent) had municipal collection and 85 (or 58 percent) had private collection. The survey was concerned specifically with the collection of garbage, which was variously defined but in general can be said to mean food waste.

The 1929 Municipal Index Survey[2]

This survey also dealt only with garbage collection. It covered 667 cities with populations over 4,500; it is not known whether or not more cities were

Table 4-1
Surveys of Organizational Arrangements for Refuse Collection

Survey	Date	Sample Size	Responses	Response Rate	Service Recipient[a]	Type of Service	Population Lower Limit
MIT	1902	N.R.[b]	161 cities	—	N.R.	garbage	28,000
Municipal Index	1929	N.R.	667 cities	—	R, C	garbage	4,500
APWA	1939	N.R.	190 cities	—	R, C	all refuse	N.R.
APWA	1955	2,390	908 cities	38%	R, C	all refuse	5,000
APWA	1964	3,117	995 cities	32%	R, C, Ins, Ind	all refuse	5,000
PHS	1968	ALL	6,259 cities	—	R, C, Ins, Ind	all refuse	5,000
Ralph Stone and Co.	1969	N.R.	234 cities	—	R, C	all refuse	10,000
AMS	1971	1,000 selected private firms	same	100%	R, C, Ind	all refuse	N.R.
Public Works	1973	5,334	1,630 cities	31%	R, C	all refuse	2,500
APWA	1973	1,300	661 cities	51%	R	all refuse	5,000
ICMA	1974	2,293	1,092 cities	48%	R, C, Ind	all refuse	10,000
Columbia University	1975	1,378[c]	1,377 cities	100%	R, C, Ins, Ind	all refuse	2,500

[a]R = Residential; C = Commercial; Ins = Institutional; Ind = Industrial
[b]N.R. = Not reported
[c]Drawn from a universe of 2,060 cities and townships

originally in the sample. Although this survey was quite thorough (it included questions as to frequency of collection and type of equipment) the information was not aggregated. Of those cities, however, 247 (or 37 percent) had some form of municipal collection, 198 (or 30 percent) had some form of contract collection, 154 (or 23 percent) had some form of private collection, and 110 (or 16 percent) reported no systematic collection at all. Although the survey was concerned with residential and commercial collection, the data as presented do not distinguish between the two.

1939 APWA Survey

The first survey by the American Public Works Association was conducted by mail in 1939.[3] It reported information from 190 cities in the United States and Canada, but it gave no indication of the number of cities that failed to respond. The reporting format used in this survey was also used in later surveys, and thus provided a base for future comparisons. Some of the terms used in the survey were carefully defined: municipal collection was defined as the case where the city pays the collection employees and the collection service is operated by a municipal department; in contract collection the city pays a contractor for doing collection work; and private collection is the case where the citizens individually, or in limited groups, pay the collectors or private collection agencies. The survey covered the collection of garbage, rubbish, and ashes (as opposed to garbage only) from residential and commercial service recipients.

It was found that 149 (or 78 percent of the cities in the sample had some form of municipal collection, 45 (or 24 percent) had some form of contract collection, and 101 (or 53 percent) had some form of private collection. The respective figures for residential collection were 78 percent, 23 percent, and 29 percent; and for commercial collection the figures were 66 percent, 22 percent, and 53 percent. The information was also given by population, with the general conclusion that larger cities were more likely to have municipal collection, and for collection of garbage only (to facilitate comparison with the 1929 survey), from which it was apparent that municipalities were even more likely to collect garbage than to collect mixed refuse, and that municipal involvement in collection had increased over that decade. No cities without systematic collection were mentioned—it would seem that these had either been eliminated by editing or else did not respond to the survey. This must be taken into account when comparing this survey with others.

1955 APWA Survey

The second APWA survey was considerably more ambitious than the first one, and attempted to cover all cities with populations of 5,000 and over in the

United States and Canada.[4] The response rate, however, was only 38 percent (908 cities). In addition, a comprehensive survey was mailed to a selected sample of 125 cities, of which 89 (71 percent) were returned. Otherwise, the type of collection covered, the format, and the definitions were largely the same as in the 1939 survey, with the exception that combined refuse, combustible and noncombustible rubbish, and garden rubbish were added as possible types of refuse. Briefly, of the 893 responses used, 72 percent showed some form of municipal collection, 28 percent some form of contract collection, and 22 percent some form of private collection. This presumably refers to collection from any type of recipient. No pattern is evident as to the use of a particular arrangement to collect a particular type of refuse. A population breakdown of the figures is available only for the detailed survey of 89 cities.

1964 APWA Survey

The next APWA survey,[5] in 1964, was similar to the preceding one. The types of service recipients covered were expanded to include manufacturing/industrial and institutional/public. Accordingly, the definition of private collection was amended to include the situation where firms and institutions (not only citizens, as in the 1939 definition) hired private hauling firms to collect their refuse. The sample again attempted to include all cities with populations of 5,000 and over, and again the response rate was less than 40 percent, as 1,116 cities replied. Of the 995 usable replies, 65 percent reported some municipal collection, 27 percent some contract collection, and 34 percent some private collection (again for any type of recipient). Apparently, the trend toward municipal collection over the previous few decades was reversed during the late fifties and early sixties. The response rate was again quite low and again biased in favor of larger cities. It is important to note that if a collector had a contract or franchise agreement with the city, and was paid by the service recipient, this arrangement was defined as "contract collection," which differs from the definition in Chapter 3. Where refuse was removed without charge, for its salvage value, it was designated private collection.

1968 PHS Survey

In 1968 the Bureau of Solid Waste Management of the U.S. Public Health Service (PHS), through the various state solid waste commissions, undertook the first government-sponsored comprehensive survey of community solid waste practices. Its initial intent was to cover all portions of Standard Metropolitan Statistical Areas (SMSAs) and all other incorporated communities of 5,000 or more population. However, only 30 states and the District of Columbia responded in time for their data to be included and an additional three states were surveyed only partially. Data were obtained from 6,259 communities with a total population of 92.5 million, of which 75 percent was urban. While the interim report[6] terms the sample coverage "not bad," some regions were covered much less thoroughly than others; for example, only 16 percent of the Southeast was covered.

The types of systems considered were public agency, private collector, and individual—where "individual" means self-service by individual householders, institutions, or industries—which do not correspond to previous surveys. Household wastes were defined as garbage, combined or separated rubbish, "bulky items," and "others." Residential, commercial, industrial and institutional service recipients were covered. Municipal collection was found (to a greater or lesser extent) in only 41 percent of the communities covered, and municipal jurisdiction over collection in 55 percent, although on a population basis these figures were 64 percent and 84 percent. On a population basis, 56 percent of households were found to have public collection, 32 percent had private collection, and 12 percent had "individual" service as defined. The respective figures for commercial wastes were 25 percent, 62 percent, and 13 percent; and for industrial wastes 13 percent, 57 percent, and 30 percent. As a final cautionary note, however, the report states that its collection figures "have [a] lower reliability [than its other figures]."

1969 Ralph Stone Survey

Ralph Stone & Company conducted a limited survey of solid waste collection (for the Public Health Service of the U.S. Department of Health, Education, and Welfare), the primary purpose of which was to define areas of possible savings with respect to crew size, but which also covered type of collection.[7] The survey covered 234 United States cities with populations greater than 10,000 (in 42 states), but as it refers at one point to the "234 responses" the size of the original sample is not clear. Collection systems were classified as municipal, private, or combined (like the PHS survey, this used cruder categories than APWA did); and types of service recipients were residential and commercial. Briefly, with respect to the 234 cities, 151 (or 64.5 percent) had municipal collection of residential refuse, 65 or (27.8 percent) had private collection, and 18 (or 7.7 percent) had mixed collection.

1971 Applied Management Sciences Survey

The private solid waste collection industry was surveyed by Applied Management Sciences, Inc.[8] Their sample included all private firms involved in the collection of solid waste in 124 (of 229) SMSAs and 51 of the 1,401 cities which had populations of 5,000 or more and were not in SMSAs. Service by private firms was divided into two categories: direct contract, where the service recipient himself arranges and pays for collection by a private firm; and government franchise, where a unit of government awards a franchise *or* contract to a private firm. The choice of terminology was confusing, in that what the APWA defined as "contract" would be classified as "government franchise" in this scheme, and what is termed "direct contract" here would be labeled "private" according to the APWA definition. Furthermore, this is not a simple transposition of terms; a franchise that calls for the service recipient to make direct payment to the private firm which provides service would be called "government franchise" in the AMS study and "private" in APWA terms.

Residential, commercial, and industrial customers were considered, although a further element of confusion was added by considering all multiple dwellings with five or more units as commercial establishments. Extrapolating from the data (from 59 percent of all private collectors) the study estimated that 50 percent of the population was served by the private sector (and therefore the other 50 percent was served either by the municipality, by self-service, or by other methods). With the inclusion of multiple dwellings in the residential category, the proportion was 55 percent private and 45 percent other. It was further estimated that private contractors collect refuse from 91 percent of the commercial and 94 percent of the industrial estalishments in the United States. Information was presented only on the fraction of the *population* serviced by private firms, not on the fraction of *cities* so serviced.

1973 Public Works *Survey*

The editors of *Public Works* magazine solicited information on the organization of solid waste collection as part of their 1973 City Engineering Data Survey.[9] Although the basic format is somewhat rudimentary—possible systems are city, private, or a combination of both—the survey is interesting because it was the first to ask about anticipated organizational changes. Also, a follow-up questionnaire was sent to those cities that indicated they were considering a change from private to municipal collection.

The survey was sent to all cities in the United States with over 2,500 population. Of those 5,334 cities, about 31 percent responded to the questions. In brief, 948 cities, or 58 percent of those reporting, had municipal collection of residential refuse, 561 or 35 percent had private collection, and 118 or 7 percent had a combination of both. Collection or commercial refuse was done by the municipality in 465 cities or 31 percent, by a private firm in 725 or 49 percent, and by a combination in 296 or 20 percent. No change in their present system of refuse collection was anticipated by 1,304 cities or 95 percent, while 42 or 3 percent were considering a change from private to municipal, and 25 or 2 percent from municipal to private.

The follow-up questionnaire was returned by 21 of the 42 cities reportedly considering a private to municipal change. Of these, five had either misunderstood the original question or had since decided against a change. Two of the remaining 16 cited dissatisfaction with their present system as the primary reason for change, five cited expected savings, four cited both of the preceding reasons, and five cited other reasons. The contemplated change was for residential collection only in three of the cities, for both residential and commercial in five cities, while the other eight cities were undecided about the scope of the change.

1973 APWA Survey

The most recent APWA survey, conducted in 1973, was carried out in much the same way as their previous surveys.[10] One difference is that this dealt only with residential collection, while their previous surveys did not specify the service recipient. The questionnaire was sent in 1973 to 1,300 cities in the United States and Canada with over 5,000 population, from which 661 usable replies were received. The response rate is thus 51 percent.

The survey found that 65 percent of all cities had at least some municipal collection, 33 percent some contract collection, and 39 percent some private collection. However, while the 1,300 cities in the sample include only 29 percent of cities over 5,000 population, they include all cities over 500,000 population. As it is already well known that larger cities are more likely to have municipal collection, the results may be biased and may tend to exaggerate the extent of municipal collection.

1974 International City Management Association (ICMA) Survey

In 1974, ICMA conducted a mail survey that was largely in the mold of previous APWA surveys, although with some significant differences.[11] All cities with populations 10,000 and over were surveyed, and 48 percent (or 1,092 cities) responded. Possible arrangements were municipal, contract, private, and combinations of those three. Although the survey appears to be comprehensive, as presented, it does not permit comparisons with other surveys.

After some additional calculations, however (see Table 4-2) one can determine that while 664 cities (60.9 precent) report that they collect residential solid waste, in only 412 cities (that is, 1,091 minus 679), or 38 percent, is the municipality the only collector of such wastes. *If* the 141 cities that report using "other systems" do not have *any* private firms collecting residential solid waste,

Table 4-2
Partial Presentation of Data from 1974 ICMA Survey

Number of Cities Reporting	Have municipally operated systems?			
	Yes		No	
1,091	664	(61%)	427	(39%)

Type of Collection System Where Municipality Does Not Collect All Residential Solid Waste.

Number of Cities	Citizens Hire Private Haulers		Municipality Contracts With Private Haulers		Other	
679	417	(61%)	291	(43%)	141	(21%)

then 538 cities (that is, 679 minus 141), or 49.3 percent have residential solid waste collection (to some extent) by private firms. At the other extreme, *if* the 141 cities that report "other systems" *all* use private firms, but are simply reporting that some of their citizens haul their own refuse to disposal sites, then all 679 cities (or 62 percent) in which the municipality does not collect all such wastes would have some private collection. (Further, the number of cities in which citizens hire private haulers, and in which municipalities contract with private haulers, adds up to more than 679, indicating that some do both.) Nevertheless, the published analysis calls attention to the 60.9 percent figure and comments that "it had previously been estimated that the mix between public and private residential systems was about 50/50." This statement could be interpreted as implying the mix is now 61/39, whereas it is not more disparate than 61/49, and could actually be 61/62, or about 50/50.

Discussion

The finding from these studies are summarized in Table 4-3, together with the findings from the Columbia University survey of 1975. (The results of the 1902 and 1971 surveys cannot be shown in this format.)

There are several intrinsic shortcomings in these surveys of organizational arrangements that make it difficult to interpret the findings and to detect changes in the prevalence of organizational arrangements. The principal defects of prior surveys, taken as a group, include the following:

Low response rates to survey questionnaires, with probable biases in the returns;

Use of mail surveys in all but one case, with ambiguous responses;

Unclear definitions of organizational arrangements;

Table 4-3
Summary of Survey Findings

A. Cities with Collection by Municipal Agencies or Private Firms

	1929 Municipal Index	1939 APWA	1955 APWA	1964 APWA	1968 PHS	1969 Stone	1973 Public Works	1973 APWA	1974 ICMA	1975 Columbia University
Percentage of cities with *any* municipal collection	44%	67%	72%	65%	41%	72%	65%	65%	61%	37.3%
Percentage of cities with *any* collection by private firms[a]	61	45	45	56	N.R.[b]	36	42	61	49-62	66.7

B. Distribution of Cities by Residential Service Arrangement, in percentages

Service Arrangement										
Municipal	39%	38%	55%	45%				39%		34%
Contract	32	4	15	18				16		19
Private	23	7	11	13				12		42[a]
Municipal and Contract	2	5	8	3				6		0.1
Municipal and Private	3	31	6	15				16		3[a]
Contract and Private	1	11	2	5				7		1[a]
Municipal, Contract and Private		5	3	2				4		0.1

[a]"Private" includes both franchise and private collection, as defined in Chapter 3.
[b]Where entries are absent, data were not reported.

Inconsistent definitions of organizational arrangements;

Inclusion or exclusion of various classes of service recipients;

Inclusion or exclusion of various classes of solid waste;

Varying limits on the sizes of cities included in the surveys;

Presentation of data in ways which defy comparison.

Therefore, only the 1969 (Stone) and 1974 (ICMA) surveys, both based on populations of 10,000 and above, could conceivably be compared with each other, and the 1955, 1964, and 1973 (APWA) surveys, based on populations of 5,000 and above, can likewise be compared with each other; all other comparisons would be suspect.

All the surveys were conducted by mail, except for the Applied Management Science survey and the Columbia University survey. The response rates of the mail surveys, where reported, were substantially less than 100 percent and, in fact, did not exceed 51 percent. One can offer the reasonable hypothesis that cities with municipal collection were more likely to respond to mail questionnaires directed to city officials than cities without municipal collection. If so, the extent of municipal collection is overstated in these surveys.

The Columbia University Survey

Learning from the defects of prior surveys, the Center for Government Studies of Columbia University conducted a survey of organizational arrangements in 1975. The survey covered residential, commercial, institutional, and individual solid waste, as well as street cleaning, litter-basket pickups, and the cleaning of parks and public places. (See Appendix A.) The definitions employed were those in Chapter 3. The survey was carried out by telephone calls to local government officials instead of by mail, to achieve the following:

1. Speed of completion of the survey;
2. High response rate, thereby minimizing error due to bias in responses;
3. Accuracy, through verbal discussion and clarification where necessary;
4. First-hand impressions by interviewers experienced in local government (primarily from the International City Management Association and Public Technology, Inc., as to reliability of the data).

A governmental jurisdiction was included in the sample if it was located in a Standard Metropolitan Statistical Area (SMSA) which lies entirely within a single state and has a total population of less than 1,500,000. Thus, 200 of the 243 SMSAs (according to the 1970 Census) were eligible. Within these SMSAs, all incorporated municipalities with populations greater than 2,500 were eligible. The largest eligible city had a population of about 750,000. In addition, all townships were eligible if they were authorized by the state to collect or arrange for the collection of solid waste, when the population for which the township had such authority was 2,500 or greater. The total number of such governments is 2,060, and these communities constitute the statistical universe for the survey. Of these, 252 are central cities and 1,808 are satellite cities or townships. All eligible communities in 41 of the 200 eligible SMSAs were included in the sample. From the remaining 159 SMSAs, all central cities, but only one half of the satellite cities, were included in the sample. The latter were then given double weight to create the complete data file upon which the analysis is based. (The validity of this procedure was demonstrated.) In summary, the total sample size consisted of 1,378 communities, the response rate was 99.9 percent (one community could not be reached by telephone despite repeated attempts), and because of the nature of the statistical sample, it can be stated with confidence that the results describe the entire universe of 2,060 communities. It is this universe that is analyzed and described in the remainder of the chapter. (The terms *city, municipality,* and *community* are used interchangeably here, and should be understood as meaning both the cities and townships described.)

Because of the size of the sample, the way the sample was selected, the high response rate, and the fact that the survey was conducted by telephone, the author considers the findings presented here to be the most authoritative to date.

Altogether, the 200 SMSAs in the sample have a total population of 67.5 million, or 33.3 percent of the 1970 U.S. population; the 2,060 communities in turn have a population of 52 million. The sample is biased, by design, toward central cities and suburbs and ignores rural areas and small communities of less than 2,500 population. This bias reflects the study's focus on policy: refuse collection is a more significant problem in urban areas than in rural areas.

Survey Findings

Service Providers

The principal service providers are municipal governments, private firms, and the service recipient himself, that is, self-service. Table 4-4 shows service providers by service recipients in terms of percentage of cities in the sample in which the indicated provider serves the indicated recipient. Totals are over 100 percent in all cases because in some cities more than one kind of service provider can be found, although this does not imply that an individual service recipient has a choice of service provider.

It is clear that collection by private firms is quite commonplace for

Table 4-4
Number and Percentage of Cities, by Service Provider, for each Type of Service Recipient

| Service Recipient | Total Number of Cities[a] | Service Provider |||||||
| | | Municipality || Private Firm || Service Recipient || Other ||
		No.	Percentage	No.	Percentage	No.	Percentage	No.	Percentage
Streets	1,591[b]	1,467	92.2	57	3.6	NA[c]		72	4.5
Parks and Public Spaces	1,644[b]	1,352	82.2	185	11.2	NA		126	7.7
Litter Baskets	1,383[b]	1,147	82.9	193	14.0	NA		55	4.0
Residential									
Bulk	1,814	920	50.7	734	40.5	439	24.2	16	0.9
Mixed Refuse	2,060	769	37.3[d]	1,342	65.1	377	18.3	23	1.1
Multiple Dwellings	1,745	645	37.0	1,213	69.5	18	1.0	8	0.4
Institutional	1,798	618	34.4	1,167	64.9	115	6.4	46	2.6
Commercial	1,992	626	31.4	1,697	80.7	84	4.2	8	0.4
Industrial	1,453	328	22.6	1,083	74.5	233	16.0	6	0.4

[a]Excludes the cities that were surveyed but responded "don't know" for service provider.
[b]Excludes the cities that were surveyed but responded "no one" for service provider.
[c]NA = Not applicable
[d]That is, 37.3% of 2,060 communities have government agencies which collect at least some of the residential mixed refuse.

residential, commercial, institutional, and industrial service. The number of cities that have *private firms* collecting at least some of their residential refuse is almost twice as large as the number that have *municipal agencies* collecting at least some of their residential refuse. For commercial, institutional, and industrial service recipients, cities with (at least some) collection by private firms outnumber cities with (at least some) municipal collection by at least two to one. Only for municipal properties—streets, parks, and litter baskets—do municipal agencies collect solid wastes in a significant majority of cities.

If one sets up a continuum between entirely municipal collection and collection entirely by private firms, the various service recipients are found to be arrayed in a well-defined and expected pattern. Municipal properties tend toward totally municipal collection, residences are more evenly distributed, and the remaining categories—institutional, commercial, and industrial—are served predominantly by private firms. In these three categories, service recipients tend to be served by private service deliverers primarily because they require specialized service (for example, daily collection of large amounts of waste) that the municipality, for various reasons, either cannot or will not provide.

This study shows that 65.1 percent of all cities in the sample have private firms collecting mixed residential refuse, and 37.3 percent have municipal agencies doing this. With the exception of the 1968 Public Health Service and the 1971 Applied Management Sciences, Inc., surveys, previous studies had significantly understated the extent of refuse collection by private firms, particularly for residential service. Most previous surveys had indicated highly variable results (no doubt due to the deficiencies noted), centering about 65 percent for the fraction of cities with municipal collection and roughly 50 percent with private collection. The various survey results are summarized in Table 4-3. The differences can be explained in part by the overreporting of municipal collection in some of the prior surveys, as discussed earlier. In addition, the fact that this sample includes all cities with populations as small as 2,500 partially explains the high proportion of collection by private firms that was found here, as smaller cities are more likely to have private collection and large cities are more likely to have collection by a municipal agency.

This sample can be analyzed for cities with populations greater than 10,000 and greater than 5,000; the results are compared in Table 4-5 with those prior surveys in which cities of these sizes can be distinguished. For cities larger than 10,000 population, the Columbia University findings are consistent with the ICMA results for collection by private firms. For cities larger than 5,000 population, the findings concerning private firms are consistent with the recent APWA survey results and suggest that contract collection became more widespread between 1955 and 1975.

There are also other useful ways to represent the role of various service providers. Table 4-6 shows one alternative, which is to weight the figures in Table 4-4 by the population in each city; that is, this table shows the percentage

Table 4-5
Comparison of Survey Findings

A. *For Cities Larger Than 5,000 in Population*

	1955 APWA	1964 APWA	1973 APWA	1975 Columbia University[a]
Percentage of cities with any municipal collection	72%	65%	65%	43.7%
Percentage of cities with any collection by private firms	45%	56%	61%	60.2%

B. *For Cities Larger than 10,000 in Population*

	1969 Stone	1974 ICMA	1975 Columbia University[a]
Percentage of cities with any municipal collection	72%	61%	49.8%
Percentage of cities with any collection by private firms	36%	49-62%	55.3%

[a]For collection of residential mixed refuse from residences other than multiple dwellings.

Table 4-6
Percentage of Population Living in Cities That Have the Indicated Service Provider, by Type of Service Recipient

Service Recipient	Service Provider			
	Municipality	Private Firm	Self-Service	Other
Streets	89.9%	1.4%	NA[b]	3.1%
Parks	81.8	4.0	NA	8.7
Litter Baskets	80.6	5.2	NA	2.6
Residential				
Bulk	62.1	29.1	17.5	1.7
Mixed Refuse	63.2[a]	47.2	12.6	1.9
Multiple Dwellings	56.6	63.3	0.4	—
Institutional	48.3	58.6	7.2	2.0
Commercial	51.6	81.4	4.6	—
Industrial	24.0	69.9	14.5	0.3

Note: Row totals exceed 100% in some cases because more than one service provider is to be found in some cities. Row totals are less than 100% for the first three rows because some cities have no regular street cleaning service, no litter basket service, etc., and because some cities responded "don't know" about these services.

[a]That is, 63.2% of the population in the sample live in a city where municipal workers collect at least some of the residential mixed refuse.

[b]NA = Not applicable

of the population which lives in cities or towns where a given type of service provider serves a given type of service recipient. As expected, because large cities are more likely to have municipal collection systems, the fraction of *population* which lives in communities with at least some municipal service is significantly higher than the fraction of *cities* with at least some municipal service (Table 4-5) for almost every class of service recipient. At least some of the residential mixed refuse is collected by the municipality in cities where 63.2 percent of the population live, by private firms where 47.2 percent live, and by self-service where 12.6 percent of the population live. The figures in this table are most meaningful for municipal properties, which are usually served by only one provider. The figures are difficult to interpret for residential and commercial recipients, as many of these are served by more than one provider. Thus, for example, if a city has both municipal and private collection of residential refuse, the total population of that city is counted in each category.

Table 4-7 shows a third useful way to present the findings. Here the original figures from Table 4-4 are weighted both by population in each city and by the percentage of this population that is served by the given provider; that is, the figures shown are absolute percentages of the total population. For the collection of residential mixed refuse, 61.9 percent of the population in this sample receive municipal service, 35.7 percent are served by private firms, and 0.9 percent practice self-service. These figures must be used with caution; the data for the fraction of the population in a community serviced by a given arrangement may be biased inasmuch as the information was obtained from municipal employees whose estimates of the magnitude of private-sector operations may be in error. A prior study that relied on private firms for its data concluded that 50 percent of the population was serviced by private firms.[1,2]

Service Arrangement

As was shown in Chapter 3, service can be provided under a variety of different arrangements. Indeed, when all service recipients are considered, no less than twenty-one different arrangements were found. These are identified in Table 4-8, following the format of Table 3-2.

Table 4-9 shows the arrangements used by different service recipients. Some

**Table 4-7
Percentage of Population Served by Indicated Service Provider, for Collection of Residential Mixed Refuse from Other Than Multiple Dwellings**

Municipality	*Private Firm*	*Self-Service*	*Other*	*Total*
61.9%	35.7%	0.9%	1.5%	100.0%

Table 4-8
Observed Service Arrangements for Solid Waste Collection, for All Service Recipients

Service Arranger	Service Provider					
	Local Municipality	County	Another Government	Special District	Private Firm	Service Recipient	Voluntary Association
Local Municipality	X	X	X	X	contract X franchise X (mandatory)	NA	X
County		X				NA	
Special District	NA	NA	NA		X	NA	
Voluntary Association		X			X	NA	X
Service Recipient	X	X	X		franchise X (nonmandatory) private X	X	X

Notes: Arrangements marked "X" were identified in the survey.
Cf. Table 3-2.

Table 4-9
Number and Percentage of Cities, by Arrangement, for Each Type of Service Recipient

Service Recipient	Total No. of Cities	Total Arrangements		Service Arrangement																	
				Municipal		Contract		Franchise		Private		Self-Service		Special District Contract		Inter-Governmental Contract		County		Municipality/Special District		Voluntary Association		Special District		Other	
	No.	No.	%	No.	%	No.	%	No.	%	No.	%	No.	%	No.	%	No.	%	No.	%	No.	%	No.	%	No.	%	No.	%
Streets	1,591[a]	1596	100.3	1467	92.2	57	3.6	NA		NA		NA		0		41	2.6	15	0.9	9	0.6	0		3	0.2	4	0.2
Parks and Public Spaces	1,644[b]	1664	101.2	1352	82.2	185	11.2	NA		NA		NA		1	0.1	17	1.0	41	2.5	42	2.6	7	0.4	9	0.5	10	0.6
Litter Baskets	1,383[c]	1395	100.9	1147	82.9	193	14.0	NA		NA		NA		0		17	1.2	11	0.8	8	0.6	9	0.6	4	0.3	6	0.4
Residential																												
Bulk	1,814	2126	117.2	920	50.7	170	9.4	67	3.7	514	28.3	439	24.2	4	0.2	1	–	2	0.1	0		8	0.4	0		1	–
Mixed Refuse	2,060	2540	123.3	769	37.3	437	21.2	151	7.3	786	38.2	377	18.3	7	0.3	4	0.2	2	0.1	1	–	1	–	1	–	5	0.2
Multiple Dwellings	1,745	1911	109.5	645	37.0	272	15.6	133	7.6	835	47.8	18	1.0	3	0.2	2	0.1	2	0.1	0		0		0		1	–
Institutional	1,798	1977	110.0	618	34.4	231	12.8	112	6.2	853	47.4	115	6.4	3	0.2	7	0.4	8	0.4	3	0.2	0		1	–	26	1.4
Commercial	1,992	2416	121.3	626	31.4	226	11.3	154	7.7	1318	66.7	84	4.2	1	–	0		2	0.1	0		0		0		5	0.2
Industrial	1,453	1671	115.0	328	22.6	109	7.5	93	6.4	902	62.1	233	16.0	1	0.1	0		2	0.1	0		0		0		3	0.2

[a]230 cities reporting that nobody sweeps the streets are not included in this total.
[b]62 cities reporting that nobody collects from parks are not included in this total.
[c]249 cities reporting that nobody empties litter baskets are not included in this total.

entries in this table are identical to those in Table 4-4 because the "municipal" and "self-service" arrangements remain unchanged. However, where the service provider is a private firm the arrangement may be contract, franchise, private, or special district contract, for example.

Municipal properties—streets, parks, and litter baskets—are serviced by contract in 4, 11, and 14 percent of cities, respectively. Mixed residential refuse is collected almost equally by municipal and private arrangements (approximately 38 percent of the cities use each), while 21 percent of cities award contracts and 7 percent award franchises for this service. Contract arrangements are more likely to be used for residential collection than for other types of collection. Self-service is used primarily for residential refuse and by industrial establishments.

It is worth noting (although it does not appear on this table) that 23 percent of cities with municipal collection of residential refuse employ different arrangements for collecting waste from institutions. In other words, it does not follow that service by a municipal agency is routinely extended to public institutions; schools and hospitals sometimes make their own separate arrangements, and do not obtain service from the municipal refuse-collection agency.

Table 4-10 shows arrangement type by service recipient for those cities in which a given recipient is served only by one arrangement. The totals in this table for municipal properties are misleadingly low. While, for instance, the 79.5 percent figure for residential mixed refuse implies that almost all the remaining 20.5 percent of the cities in the sample are served by two or more arrangements, the 63.9 percent figure for litter baskets should be taken to mean that in most of the remaining cities, there is *no* collection from litter baskets (and, presumably, there are no litter baskets).

It is useful to look more closely at the number of cities which utilize each arrangement, and the different combinations of arrangements, for residential service. Table 4-11 does so, for all cities in the sample, for cities larger than 5,000 in population, and for cities larger than 10,000. A very revealing statistic that can be derived from the data here is that substantially more communities (46 percent of all cities) rely entirely on private firms for service provision than on municipal agencies (32.3 percent of all cities). Cities with any degree of self-service are not considered to be reliant exclusively on municipal or private service providers, in this context. However, if we look at cities in terms of their reliance on the public or private sectors for organized service, whether or not there is supplemental self-service as well, we find that 61.4 percent of all cities rely on private firms, whereas only 33.6 percent of all cities rely on municipal collection.

The data for cities larger than 5,000 can be compared, in Table 4-12, with data from the prior APWA surveys. With respect to municipal collection, the 1975 Columbia University survey is in agreement with the 1973 APWA survey and provides supporting evidence that between 1955 and 1975 there was a

Table 4-10
Number and Percentage of Cities in Which a Class of Service Recipient Is Served by Only a Single Arrangement

Service Recipient	Total		Municipal		Contract		Franchise		Private		Self-Service	
	No.	Percentage	No.	Percentage	No.	Percentage	No.	Percentage	No.	Percentage	No.	Percentage
Streets	1515	73.5	1462	71.0	53	2.6	NA	NA	NA	NA	NA	NA
Parks and Public Spaces	1506	73.1	1333	64.7	173	8.4	NA	NA	NA	NA	NA	NA
Litter Baskets	1316	63.9	1131	54.9	185	8.9	NA	NA	NA	NA	NA	NA
Residential												
Bulk	1489	72.3	766	37.2	150	7.3	52	2.5	291	14.1	230	11.2
Mixed Refuse	1637	79.5	666	32.3	394	19.1	99	4.8	435	21.1	10	.5
Multiple Dwellings	1569	76.2	511	24.8	241	11.7	130	6.3	777	32.9	10	.5
Institutional	1580	76.7	503	24.4	201	9.8	107	5.2	708	34.4	61	3.0
Commercial	1567	76.1	353	17.1	136	6.6	135	6.6	935	45.4	8	.4
Industrial	1235	60.0	224	10.9	78	3.9	84	4.1	742	36.0	107	5.2

Table 4-11
Distribution of Cities by Arrangement and Combinations of Arrangements for Collection of Mixed Refuse from Residences Other than Multiple Dwellings

Arrangement	All Cities		Cities Over 5,000 pop.		Cities Over 10,000 pop.	
	No.	Percentage	No.	Percentage	No.	Percentage
Municipal only	666	32.3	512	37.8	358	42.9
Contract only	394	19.1	266	19.6	158	18.9
Franchise only	99	4.8	71	5.2	53	6.3
Private only	435	21.1	219	16.2	95	11.4
Self-Service only	10	0.5	5	0.4	2	0.2
Only one other	3	0.1	2	0.1	1	0.1
Municipal and Contract	3	0.1	3	0.2	3	0.4
Municipal and Franchise	3	0.1	0	0	0	0
Municipal and private	52	2.5	43	3.2	35	4.2
Municipal and Self-Service	27	1.3	22	1.6	10	1.2
Municipal and Other	7	0.3	2	0.1	2	0.2
Contract and Private	18	0.9	14	1.0	13	1.6
Contract and Self-Service	16	0.8	10	0.7	10	1.2
Franchise and Private	3	0.1	3	0.2	3	0.4
Franchise and Self-Service	46	2.2	23	1.8	19	2.3
Private and Self-Service	254	12.3	141	10.4	61	7.3
Private and Other	4	0.2	1	0.1	1	0.1
Self-Service and Other	5	0.2	1	0.1	1	0.1
Municipal, Contract, and Private	3	0.1	3	0.2	3	0.4
Municipal, Private, and Self-Service	9	0.4	8	0.6	4	0.5
Contract, Private and Self-Service	3	0.1	2	0.1	1	0.1
Franchise, Private, and Self-Service	1	0	1	0.1	1	0.1
Private, Self-Service, and Other	4	0.2	2	0.1	1	0.1
Total	2,060	99.7	1,356	99.8	835	100.0

decline in the fraction of cities utilizing only municipal collection for residential refuse. Given the agreement between the two surveys for municipal collection, the higher level of contract and private collection found in 1975 compared to 1973 supports the hypothesis that the APWA survey method was likely to underreport the number of cities that have only contract or only private collection.

The data shown for the Columbia University survey in Table 4-12 are for collection of mixed refuse from residences which are other than multiple

Table 4-12
Comparison of Survey Results for Cities Larger than 5,000 in Population

Service Arrangement	1955 APWA	1964 APWA	1973 APWA	1975 Columbia University[a]
Municipal	55%	45%	39%	38%
Contract	15	18	16	20
Private[b]	11	13	12	21
Municipal and Contract	8	3	6	0.2
Municipal and Private[b]	6	15	16	3.2
Contract and Private[b]	2	5	7	1.2
Municipal, Contract, and Private[b]	3	2	4	0.2
Total	100	101	100	64[c]

[a] For collection of mixed refuse from residences other than multiple dwellings.
[b] "Private" includes both franchise and private collection.
[c] The other 36 percent of cities use self-service, either alone or, usually, in combination with one or more organized collection arrangements. It is not clear how the APWA surveys treated such cities.

dwellings, however the latter are defined locally. Therefore, the higher levels of multiple arrangements found in the APWA surveys may be due to the inclusion therein of data for multiple dwellings, which in any city are often serviced by an arrangement other than the one used to service small residences.

Just as different service recipients are associated with different arrangements, various demographic factors are also associated with different arrangements. Table 4-13 shows that cities with larger populations are considerably more likely to have municipal service for residential collection, and the larger the city, generally, the more likely is this the case. Figure 4-1 shows this graphically.

When the data are examined by region, some sharp differences emerge. Municipal collection predominates in the South by a wide margin, and is far more likely to be found there than elsewhere, while franchise collection is much more common in the West than in other parts of the country. Northern cities in the East and Midwest use private collection more than other arrangements, and private collection is more common there than elsewhere. The different forms of local government are quite similar in their use of the various collection arrangements, except for the relatively rare town-meeting form of government; that is, mayor-council and council-manager cities show similar patterns of use of the different arrangements.

Table 4-13
Service Arrangements for Collection of Residential Mixed Refuse

	Total		Municipal		Contract		Franchise		Private		Self-Service		Other	
	No.	Percentage	No.	Percentage	No.	Percentage	No.	Percentage	No.	Percentage	No.	Percentage	No.	Percentage
Total	2,540	99.9	769	30.3	437	17.2	151	5.9	786	30.9	377	14.8	20	0.8
Population Group														
>500,000	15	100.0	8	53.3	1	6.7	1	6.7	2	13.3	1	6.7	2	13.3
250,000-499,999	25	100.0	18	72.0	1	4.0	1	4.0	3	12.0	1	4.0	1	4.0
100,000-249,999	97	100.0	63	64.9	8	8.2	2	2.1	15	15.5	7	7.2	2	2.1
50,000-99,999	175	100.0	87	49.7	23	13.1	15	8.6	28	16.0	21	12.0	1	0.6
25,000-49,999	204	99.9	64	31.3	42	20.6	18	8.8	53	26.0	27	13.2	0	
10,000-24,999	503	100.1	179	35.6	113	22.5	39	7.8	117	23.3	54	10.7	1	0.2
5,000-9,999	640	100.0	178	27.8	110	17.2	24	3.8	219	34.2	107	16.7	2	0.3
2,500-4,999	881	100.0	172	19.6	139	15.8	51	5.8	349	39.6	158	18.0	11	1.2
Geographic Region														
Northeast	983	100.0	186	18.9	218	22.2	18	1.8	383	39.0	176	17.9	2	0.2
North Central	720	100.0	142	19.7	98	13.6	30	4.2	332	46.1	108	15.0	10	1.4
South	470	100.0	343	73.0	48	10.2	14	3.0	33	7.0	27	5.7	5	1.1
West	367	100.0	98	26.7	73	19.9	89	24.3	38	10.3	66	18.0	3	0.8
Metro/City Type														
Central	313	99.9	192	61.3	28	8.9	17	5.4	46	14.7	24	7.7	6	1.9
Suburban	2,227	99.9	577	25.9	409	18.4	134	6.0	740	33.2	353	15.8	14	0.6
Form of Government														
Mayor-Council	882	100.0	373	42.3	212	24.0	45	5.1	181	20.5	65	7.4	6	0.7
Council-Manager	726	100.0	321	44.2	124	17.1	88	12.1	101	13.9	87	12.0	5	0.7
Commission	67	100.0	34	50.7	15	22.4	2	3.0	9	13.4	4	6.0	3	4.5
Town Meeting	112	100.0	16	14.3	16	14.3	2	1.8	43	38.4	35	31.2	0	
Rep. Town Meeting	20	100.0	4	20.0	2	10.0	0		8	40.0	6	30.0	0	
Don't Know	733	100.1	21	2.9	68	9.3	14	1.9	444	60.6	180	24.6	6	0.8

Note: This table shows the distribution of *arrangements*, not the distribution of *cities*. The total number of arrangements is 2,540 in the 2,060 cities. See Appendix A for list of states in each region.

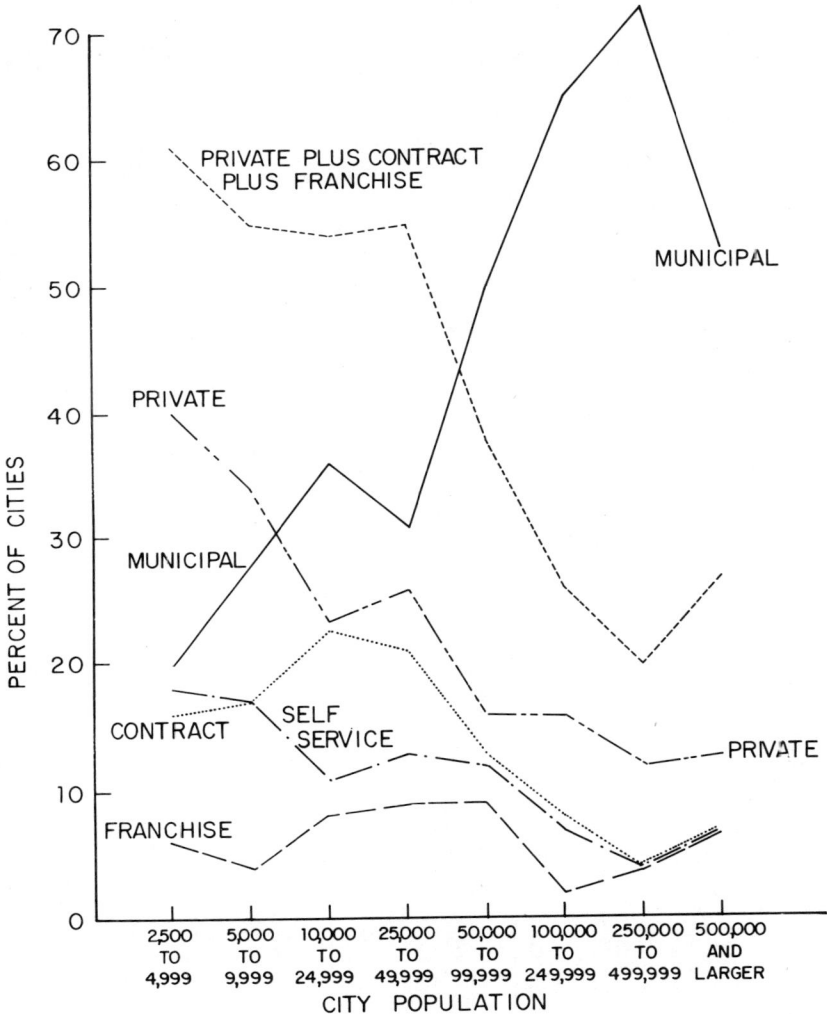

Figure 4-1. Arrangements Used for Residential Mixed Refuse Collection by City Size

Types of Residential Service

There are other, more specialized types of residential service that bear mentioning. Table 4-14 shows, for instance, that in 134 cities (6.5 percent of the sample) wet garbage is collected separately, primarily by the municipality.

Table 4-14
Arrangements for Other Types of Residential Service

Service Type	Total		Municipal		Contract		Franchise		Private		Self-Service	
	No.	Percentage	No.	Percentage	No.	Percentage	No.	Percentage	No.	Percentage	No.	Percentage
Wet Garbage	134	99.9	64	47.8	42	31.3	3	2.2	18	13.4	7	5.2
Yard Trash	470	99.9	354	75.3	38	8.1	20	4.2	23	4.9	33	7.0
Newspapers	58	99.9	20	34.5	9	15.5	1	1.7	1	1.7	5	8.6

Yard trash is collected separately in 470 cities (22.8 percent of the sample), and newspapers in 58 cities (2.8 percent of the sample). The latter figure no doubt substantially understates the extent of newspaper collection and recycling in the country, largely because such collection usually does not fall under the purview of the agency or firm concerned with the collection of most other residential solid wastes.

Changes in Arrangements

During the five-year period 1970 to 1974, government and private "shares of market" remained approximately constant for cities larger than 25,000 in population. In smaller cities, however, government collection of residential refuse increased at the expense of private-sector collection, as measured by the number of cities using the different arrangements. (See Table 4-15.) A total of sixty cities in the sample reported changing from a private to a municipal service provider, while thirty-seven reported changing in the other direction. In addition, three cities in the South changed from private to county collection. Eleven previously "self-service" cities in the sample began organized collection, nine with private service providers and two with municipal. During this time, twenty-one other cities changed collection arrangements, primarily within the private sector (from private to contract collection, or vice versa). Overall, 129 cities, or 6.3 percent of the sample, reported an organizational change. This is equivalent to a yearly rate of about 1.25 percent; that is, one out of every eighty cities changed its collection arrangement during the course of a year. One out of 106 cities changed from a private to a municipal service provider, or vice versa.

Note that there are effects associated with region and form of government. Northern cities shifted slightly toward municipal collection but southern and western cities showed no trend. Cities with the mayor-council form of government tended to change to municipal collection, while those with the council-manager form showed no preferred direction of change.

The finding here that more cities changed *to* rather than *from* municipal collection in 1970-1975 fails to confirm the indication that municipal collection declined in the period 1955-1975. Future studies, using comparable methods, will be needed to monitor subsequent shifts.

Extrapolation of Findings

The 200 SMSAs in the sample have a total population of 67.5 million, and the 2,060 communities a population of 52 million. The original sample included only ten of the twenty-five largest cities in the country. In an effort to extend the applicability of the findings, a supplementary survey of the remaining fifteen large cities was conducted and the results were compared with the ten large cities in the original sample (Table 4-16). No major differences were found.

Table 4-15
Changes in Residential Service Arrangements, in Terms of Service Providers[a]

	Total		Private Firm to Municipal		Municipal to Private Firm		Self-Service to Municipal		Self-Service To Private Firm		Other	
	No.	Percentage	No.	Percentage	No.	Percentage	No.	Percentage	No.	Percentage	No.	Percentage
Total	129	100.1	60	46.5	37	28.7	2	1.6	9	7.0	21	16.3
Population Group												
>500,000	0		0		0		0		0		0	
250,000-499,999	1	100.0	1	100.0	0		0		0		0	
100,000-249,999	4	100.0	2	50.0	2	50.0	0		0		0	
50,000-99,999	3	100.0	2	66.7	0		0		0		1	33.3
25,000-49,999	14	100.0	5	35.7	5	35.7	0		0		4	28.6
10,000-24,999	36	100.0	21	58.3	10	27.8	1	2.8	4	11.1	0	
5,000-9,999	26	100.0	11	42.3	11	42.3	0		0		4	15.4
2,500-4,999	45	100.0	20	44.4	7	15.6	1	2.2	5	11.1	12	26.7
Geographic Region												
Northeast	48	100.1	26	54.2	8	16.7	0		6	12.5	8	16.7
North Central	36	100.1	15	41.7	9	25.0	2	5.6	2	5.6	8	22.2
South	22	99.9	9	40.9	10	45.4	0		0		3	13.6
West	23	100.0	10	43.5	10	43.5	0		1	4.3	2	8.7
Metro/City Type												
Central	11	99.9	5	45.4	5	45.4	0		0		1	9.1
Suburban	118	99.9	55	46.6	32	27.1	2	1.7	9	7.6	20	16.9
Form of Government												
Mayor-Council	67	100.0	38	56.7	19	28.4	1	1.5	2	3.0	7	10.4
Council-Manager	32	100.0	13	40.6	15	46.9	0		3	9.4	1	3.1
Commission	2	100.0	0		0		0		0		2	100.0
Town Meeting	3	100.0	3	100.0	0		0		0		0	
Rep. Town Meeting	1	100.0	0		1	100.0	0		0		0	
Don't Know	24	100.0	6	25.0	2	8.3	1	4.2	4	16.7	11	45.8

[a] During the five-year period 1970-1974. See Appendix A for list of states in each region.

Table 4-16
Service Arrangements for Large Cities

A. For the Ten Largest Cities in the Sample

Service Recipient	Service Arrangement					
	Municipal %	Contract %	Franchise %	Private %	Self-Service %	Other %
Parks and Public Spaces	80[a]	0	0	0	0	30
Residential						
Mixed Refuse	80	10	10	20	10	20
Multiple Dwellings	70	20	30	40	0	0
Commercial	50	30	80	70	0	0
Industrial	0	10	30	70	10	0

B. For the Other Fifteen of the Twenty-Five Largest Cities in the United States

Parks and Public Spaces	93.3	6.7	0	0	0	6.7
Residential						
Mixed Refuse	86.7	6.7	6.7	0	0	0
Multiple Dwellings	46.7	0	6.7	60.0	0	0
Commercial	53.3	0	6.7	86.7	0	0
Industrial	26.7	0	6.7	93.3	6.7	0

[a]That is, 80 percent of the cities have municipal collection of refuse from parks and open spaces.

Table 4-17 shows the arrangements used for service delivery in the suburban cities located in the SMSAs that contain the ten largest cities in the sample. The arrangements for collecting mixed residential refuse in those cities are very similar to the arrangements in all the suburban communities in the entire set of 200 SMSAs, as shown in Table 4-13.

In other words: the largest cities in the country, which are the only urban areas with significantly different demographic characteristics than those covered by the sample, have organizational arrangements for solid waste collection that are generally similar to those in the largest cities in the sample; and the SMSAs containing the large cities in the sample have suburbs that are much more similar (in terms of organizational arrangements for solid waste collection) to all suburbs in the sample than to the central cities of the SMSAs. Therefore, it seems reasonable to extrapolate the findings of this chapter to all cities in all SMSAs.

The logic of this extrapolation can be visualized in Table 4-18. A, A', and B were found for the indicated cities, and therefore B can reasonably be extended (shown by B') to the suburban cities in the SMSAs—not in the sample—that contain the largest cities in the country. Furthermore, there is no reason to believe that the other excluded SMSAs, the interstate ones, differ from those in

Table 4-17
Service Arrangements by Service Recipient for Suburban Cities in SMSAs with the Ten Largest Cities in the Sample

Service Recipient	Total		Service Arrangement									
			Municipal		Contract		Franchise		Private		Self-Service		Other	
	No.	Percentage	No.	Percentage	No.	Percentage	No.	Percentage	No.	Percentage	No.	Percentage	No.	Percentage
Street Sweepings	166	100.0	155	93.4	7	4.2	NA	NA	NA	NA	NA	NA	4	2.4
Litter Baskets	128	100.0	100	78.1	23	18.0	NA	NA	NA	NA	NA	NA	5	3.9
Parks and Public Spaces	56	100.0	114	73.1	18	11.5	NA	NA	NA	NA	NA	NA	24	15.4
Residential														
Bulk	225	100.0	81	36.0	11	4.9	13	5.8	59	26.2	56	24.9	5	2.2
Mixed Refuse	252	100.0	71	28.2	38	15.1	19	7.5	68	27.0	50	19.8	6	2.4
Multiple Dwellings	176	100.1	58	33.0	20	11.4	19	10.8	78	44.3	1	0.6	0	
Institutional	176	100.1	50	28.4	20	11.4	13	7.4	89	50.6	4	2.3	0	
Commercial	234	99.9	53	22.6	15	6.4	19	8.1	139	59.4	8	3.4	0	
Industrial	144	100.0	19	13.2	17	11.8	8	5.6	84	58.3	16	11.1	0	

Table 4-18
Logic of Extrapolation of Findings

	All SMSAs in Sample	SMSAs (in sample) with Large Cities	SMSAs (not in sample) with Large Cities	Other SMSAs not in sample (interstate SMSAs)	Other cities not in SMSAs (and not in sample)
Central Cities	A	A'	A'	A"	Not Applicable
Other Cities	B	B	B'	B"	B'''

the sample or from the large SMSAs, and so the findings are presumed to be extrapolatable to those SMSAs as well, as A" and B". The solid waste collection practices reported in this chapter could therefore be said to characterize practices for the 107.9 million persons (53 percent of all the 1970 population of the United States) living within SMSAs in places with populations exceeding 2,500.

Going one final step, there is no evident reason to believe that cities *not* in SMSAs will differ from cities (of comparable size) *in* SMSAs. Accordingly, it appears reasonable to extrapolate the findings to all places of population greater than 2,500 (to B''' in Table 4-18), and to say that the findings presented in this chapter describe the refuse-collection services of 133.5 million people, or 66 percent of the 1970 population of the United States.

Summary

This chapter describes the results of what is arguably the most complete survey to date of the different organizational arrangements used by localities for solid waste collection. Prior attempts to measure the prevalence of the different arrangements suffered from a variety of defects. By using telephone interviews, experienced interviewers, and clearer definitions, and with an almost-perfect response rate, the Columbia University survey is likely to be more accurate than earlier ones.

It was found that private firms collect residential refuse in almost twice as many cities (67 percent) as municipal agencies (37 percent) and they collect commercial, institutional, and industrial refuse in two or three times as many cities as municipal agencies. Only municipal properties—streets, parks, litter baskets—are cleaned predominantly by public agencies. Substantially more communities (61.4 percent of all cities) rely entirely on private firms rather than on municipal agencies (33.6 percent of all cities) for organized residential refuse collection.

In terms of people served, however, a different picture emerges. Because larger cities are more likely to have municipal collection, it was found that 64.1 percent of the population in the sample lives in cities where at least some of the residential mixed refuse is collected by municipal agencies, and 51 percent lives in cities where at least some is collected by private firms. About 62 percent of the population in the sample receives municipal service, 36 percent is served by private firms, and 0.9 percent practice self-service.

Twenty-one different organizational arrangements for solid waste collection were found. For residential mixed refuse collection the following principal arrangements were found: private, 38.2 percent; municipal, 37.3 percent; contract, 21.2 percent; franchise, 7.3 percent. Municipal collection is common in large cities and in the South. Private collection is common in northern cities in the East and Midwest, and in smaller cities. Franchise collection is more likely to be found in the West than elsewhere. Contract collection is widely distributed and is as likely to be found in cities with the council-manager form of government as in cities with the mayor-council form.

In the course of a year, an average of one community out of eighty changed from one arrangement to another for residential collection. One out of 106 changed from the private to the public sector, or vice versa. There is contradictory evidence as to a trend toward or away from municipal collection.

A total of 2,060 cities are represented in the survey; they have a combined population of 52 million. With reasonable extrapolation, the results can be said to describe the solid waste collection services for 133.5 million Americans, or 66 percent of the 1970 population.

Notes

1. William F. Morse, *The Collection and Disposal of Municipal Waste* (New York: Municipal Journal and Engineer, 1908).

2. American Public Works Association, *Refuse Collection Practice* (Chicago: Public Administration Service, 1941).

3. Ibid.

4. American Public Works Association, *Refuse Collection Practice*, 2nd ed. (Chicago: Public Administration Service, 1958).

5. American Public Works Association, *Refuse Collection Practice*, 3rd ed. (Chicago: Public Administration Service, 1966).

6. Anton J. Muhich, Albert J. Klee, and Paul W. Britton, *1968 National Survey of Community Solid Waste Practices: Preliminary Data Analysis*, Solid Waste Management Series SW-3s (Cincinnati: U.S. Department of Health, Education, and Welfare, 1968).

7. Ralph Stone and Company, Inc., *A Study of Solid Waste Collection Systems Comparing One-Man with Multi-Man Crews*, Solid Waste Management Series SW-9c (Washington: U.S. Department of Health, Education, and Welfare, 1969).

8. Applied Management Sciences, Inc., *The Private Sector in Solid Waste Management: A Profile of Its Resources and Contributions to Collection and Disposal*, Solid Waste Management Series SW-51d.1, 2 vols. (Washington: U.S. Environmental Protection Agency, 1973).

9. "Trends in Municipal Solid Waste Management," *Public Works* 105, no. 5 (May 1974): 68-69.

10. American Public Works Association, *Solid Waste Collection Practice*, 4th ed. (Chicago: American Public Works Association, 1975).

11. Robert J. Bartolotta, "Local Government Solid Waste Programs," *Municipal Year Book* 42 (1975): 232-241.

12. Applied Management Sciences, *Private Sector in Solid Waste Management*.

5 Service Levels for Residential Refuse Collection

The level of service in residential solid waste collection can be described in terms of frequency of collection and location of pickup. These are the principal components, and they were measured in the Columbia University survey.

Table 5-1 shows the frequency of collection by city size, region, city type, form of government, and arrangement. It should be remembered that service levels in a community may or may not be uniform for all residents. Even under municipal or contract collection, individual residents may have a choice as to the level of service they receive and for which they are willing to pay. Conversely, even under private collection an individual resident may have no choice, if no private firm is willing to provide the level of service he desires. Accordingly, cities whose collection arrangements lead to more than one collection frequency are tabulated more than once in Table 5-1.

Frequency of Service

Of the cities in the sample, a service frequency of no more than once a week is found in 58 percent, while only 32 percent have a frequency of twice per week or greater. There is no clear relation between city size and collection frequency, but central cities are more likely to have frequent collection than suburban cities, which presumably reflects the need for more frequent collection in cities with greater population densities.

As expected, southern cities have substantially more frequent collection than northern cities, no doubt due to the accelerated decay of garbage in warmer climates. Cities with municipal collection were more likely to have frequent service than cities with private service providers, whether under contract, franchise, or private collection. While council-manager cities had a somewhat higher frequency of service than mayor-council cities, no profound significance can be attributed to that fact.

Pickup Location

As Table 5-2 shows, curb or alley service is substantially more common than backyard or frontyard service: 53.1 percent of cities have the former compared to only 11.9 percent with the latter. Both pickup locations are used, however, in 33.2 percent of cities.

Table 5-1
Residential Service Level: Frequency of Collection

	Total		More than Twice/Week		Twice/Week		Once/Week		Less than Once/Week		Various	
	No.	Percentage	No.	Percentage	No.	Percentage	No.	Percentage	No.	Percentage	No.	Percentage
Total	2091	99.8	33	1.6	643	30.8	1196	57.2	17	0.8	202	9.6
Population Group												
>500,000	14	99.9	1	7.1	5	35.7	8	57.1	0		0	
250,000-499,999	24	100.0	0		15	62.5	6	25.0	0		3	12.5
100,000-249,999	86	100.0	4	4.7	37	43.0	38	44.2	0		7	8.1
50,000-99,999	149	100.0	1	.7	57	38.3	75	50.3	0		16	10.7
25,000-49,999	172	100.1	2	1.2	44	25.6	113	65.7	2	1.2	11	6.4
10,000-24,999	439	100.0	5	1.1	161	36.7	230	52.4	0		43	9.8
5,000-9,999	513	100.0	11	2.1	154	30.0	297	57.9	5	1.0	46	9.0
2,500-4,999	694	100.2	9	1.3	170	24.6	427	61.9	10	1.4	76	11.0
Geographic Region												
Northeast	776	99.9	15	1.9	178	22.9	495	63.8	10	1.3	78	10.0
North Central	583	99.9	2	0.3	69	11.8	441	75.6	7	1.2	64	11.0
South	440	99.9	15	3.4	328	74.5	71	16.1	0		26	5.9
West	292	99.9	1	0.3	68	23.3	189	64.7	0		34	11.6
Metro/City Type												
Central	277	100.1	5	1.8	124	44.8	124	44.8	1	0.4	23	8.3
Suburban	1814	99.8	28	1.5	519	28.6	1196	65.9	16	0.9	179	9.9

Form of Government										
Mayor-Council	802	100.0	15	1.9	275	34.3	443	55.2	11	1.4	58	7.2
Council-Manager	622	100.0	13	2.1	273	43.9	276	44.4	0		60	9.6
Commission	59	100.0	1	1.7	42	71.2	13	20.0	0		3	5.1
Town Meeting	68	100.0	0		7	10.3	55	80.9	2	2.9	4	5.9
Rep. Town Meeting	14	100.0	2	14.3	3	21.4	7	50.0	0		2	14.3
Don't Know	526	100.0	2	0.4	43	8.2	402	76.4	4	0.8	75	14.2
Arrangement										
Municipal	769	100.0	20	2.6	355	46.2	363	47.2	8	1.0	23	3.0
Contract	436	99.9	7	1.6	147	33.7	249	57.1	5	1.1	28	6.4
Franchise	151	99.9	0		29	19.2	105	69.5	0		17	11.2
Private	711	100.0	6	0.8	105	14.8	468	65.8	4	0.6	128	18.0
Other	19	99.9	2		7	36.8	10	52.6	0		2	10.5

Note: "Various" includes "don't know," cities where the frequency varies seasonally, and cities with multiple collection frequencies where one or more frequencies were not clearly identified by the respondent.

Table 5-2
Residential Service Level: Location of Pickup

	Total		Backyard or Frontyard		Curbside or Alley		Various		Don't Know	
	No.	Percentage	No.	Percentage	No.	Percentage	No.	Percentage	No.	Percentage
Total	2084	99.9	248	11.9	1107	53.1	693	33.2	36	1.7
Population Group										
>500,000	14	100.0	0		4	28.6	10	71.4	0	
250,000-499,999	24	99.9	5	20.8	6	26.0	11	45.8	2	8.3
100,000-249,999	86	99.9	16	18.6	34	39.5	34	39.5	2	2.3
50,000-99,999	150	99.9	26	17.3	68	42.0	59	39.3	2	1.3
25,000-49,999	170	100.0	14	8.2	95	55.9	59	34.7	2	1.2
10,000-24,999	437	100.0	56	12.8	226	51.7	144	33.0	11	2.5
5,000-9,999	511	100.0	65	12.7	282	55.2	157	30.7	7	1.4
2,500-4,999	692	99.9	66	9.5	397	53.4	219	31.6	10	1.4
Geographic Region										
Northeast	770	100.0	76	9.9	494	64.2	185	24.0	15	1.9
North Central	582	100.0	54	9.3	302	51.9	218	37.4	8	1.4
South	440	99.9	96	21.8	178	40.4	155	35.2	11	2.5
West	292	99.9	22	7.5	133	45.5	135	46.2	2	0.7
Metro/City Type										
Central	278	100.0	43	15.5	116	41.7	113	40.6	6	2.2
Suburban	1806	100.0	205	11.3	991	54.9	580	32.1	30	1.7

	N	%	N	%	N	%	N	%	N	%
Form of Government										
Mayor-Council	803	100.0	103	12.8	414	51.6	278	34.6	8	1.0
Council-Manager	621	99.9	95	15.3	271	43.6	241	38.8	14	2.2
Commission	58	100.0	9	15.5	34	55.6	15	25.9	0	
Town Meeting	66	99.9	7	10.6	44	66.6	15	22.7	0	
Rep. Town Meeting	14	99.9	0		8	57.1	6	42.8	0	
Don't Know	522	100.0	34	6.5	336	64.4	138	26.4	14	2.7
Arrangement										
Municipal	768	100.0	129	15.6	387	50.4	261	34.0	0	
Contract	435	100.0	37	8.5	307	70.6	91	20.9	0	
Franchise	151	99.9	23	15.2	60	39.7	66	43.7	2	1.3
Private	709	100.0	68	9.6	339	47.8	268	37.8	34	4.8
Other	19	100.0	0		14	73.7	5	26.3	0	

See Appendix A for list of states in each region.

It appears, contrary to intuition, that the larger the city, the more likely it is to have service at the front- or backyard rather than at the curb or alley. Southern cities are twice as likely as other cities to have backyard or frontyard collection. However, so many cities report "various pickup locations" that any generalizations about these observations are subject to serious error.

As discussed in Chapter 7, curb or alley service costs less to provide than collection from the front- or backyard. One might therefore expect that yard collection, which is more convenient for the household, would be associated with family income. Table 5-3 shows that this is indeed the case: families with incomes greater than $14,000 per year are more likely to have frontyard or backyard collection than families with lower incomes.

Excessive Service Levels

It might be expected that the level of service would be higher in cities with municipal collection as a result of three factors:

1. Bureaucratic pressure for larger agency budgets;[1]
2. Pressure by patronage-seeking politicians, by unionized employees, and by full-employment programs for an enlarged work force;
3. Overreaction by elected officials to the requests of a few citizens, resulting in a gradual upward "creeping" of service levels, without explicit cognizance of the concomitant cost increase.

Peterson argues that cities commit themselves to public expenditure levels that exceed those which taxpayer-consumers would freely choose.[2] Is it true that government agencies overproduce, that is, that they provide service that is excessive in comparison with public demand? By examining the frequency of residential refuse collection, this study casts some light on this issue.

Presumably, a municipality is subject to greatest pressure and exercises greatest discretion when it operates its own system, to somewhat less pressure

Table 5-3
Location of Pickup by Mean Annual Family Income, in Number and Percentage of Cities

Income	Location of Pickup					
	Curb or Alley		Backyard or Frontyard		Total	
	No.	Percentage	No.	Percentage	No.	Percentage
Less than $14,000	1631	81.7	366	18.3	1997	100.0
$14,000 or More	132	67.0	65	33.0	197	100.0

under a contract, still less under a franchise, and almost none under private collection. Conversely, taxpayer-consumers are best able to express and satisfy their individual needs under private collection, somewhat less under franchise collection, still less under contract collection, and least under municipal collection, where—it is shown in subsequent chapters—the full, true cost of service tends to be obscured.

Table 5-4(A), which is derived from Table 5-1, shows that cities with municipal collection are indeed more likely than are cities with contract, franchise, or private arrangements to collect residential refuse two or more times per week. Of cities with municipal arrangements, 50.8 percent had such service; totals for the other arrangements are 38.2 percent, 21.6 percent, and 19.1 percent, respectively. In other words, the frequency of collection decreases with the level of municipal involvement, a finding that tends to confirm the hypothesis that municipalities overproduce this service.

This finding is supported by examination of a smaller sample that was studied in depth in connection with the economic analysis reported in a subsequent chapter. Table 5-4(B) shows a pattern quite similar to that in the larger sample.

One wonders if this apparent excessive frequency of municipal collection may be due to city size; that is, large cities tend to have municipal collection (as shown in Figure 4-1) and they may require more frequent service for reasons of public health. Figure 5-1 demonstrates that this is not the case; what may be termed municipal over-production occurs in all size classes of cities. Within each category of city size, cities with municipal collection have a higher frequency of service than other cities, and the frequency generally declines as one looks in sequence at municipal, contract, franchise, and private collection.

Figure 5-2 reveals that municipal collection is generally provided at a higher frequency than other arrangements even when the data are controlled for population density. When the data for each region of the country are examined, separately, however, as in Figure 5-3, the evidence is mixed.

With respect to the other aspect of service level studied here, location of pickup, the data show no clear picture. Enough cities reported "various pickup locations" to make any analysis questionable.

In summary, while this evidence is not conclusive, it is very suggestive. The analysis provides empirical support for the contention that city agencies produce more public goods than the public is prepared to accept. However, one might interpret these findings as supportive of the counterhypothesis that the public demands a higher level of service than the private sector will supply, and only government is sensitive to these preferences and is willing to satisfy them. This counterhypothesis invites skepticism because there is no obvious reason why private firms would fail to provide the desired level of this service.

No one would argue seriously with the assertion that the public, or an individual resident, is entitled to whatever high level of service he or she is

Table 5-4
Frequency of Residential Collection, by Arrangement

A. For Survey Sample

Frequency	Service Arrangement							
	Municipal		Contract		Franchise		Private	
	No.	Percentage	No.	Percentage	No.	Percentage	No.	Percentage
At least twice per week	375	50.8	154	38.2	29	21.6	111	19.1
Once per week	363	49.2	249	61.8	105	78.4	468	80.8
Total	738	100.0	403	100.0	134	100.0	579	100.0
B. For Detailed Study Sample								
At least twice per week	55	56.2	38	43.1	11	19.0	23	28.4
Once per week	43	43.9	50	56.8	47	81.0	58	71.6
Total	98	100.1	98	99.9	58	100.0	81	100.0

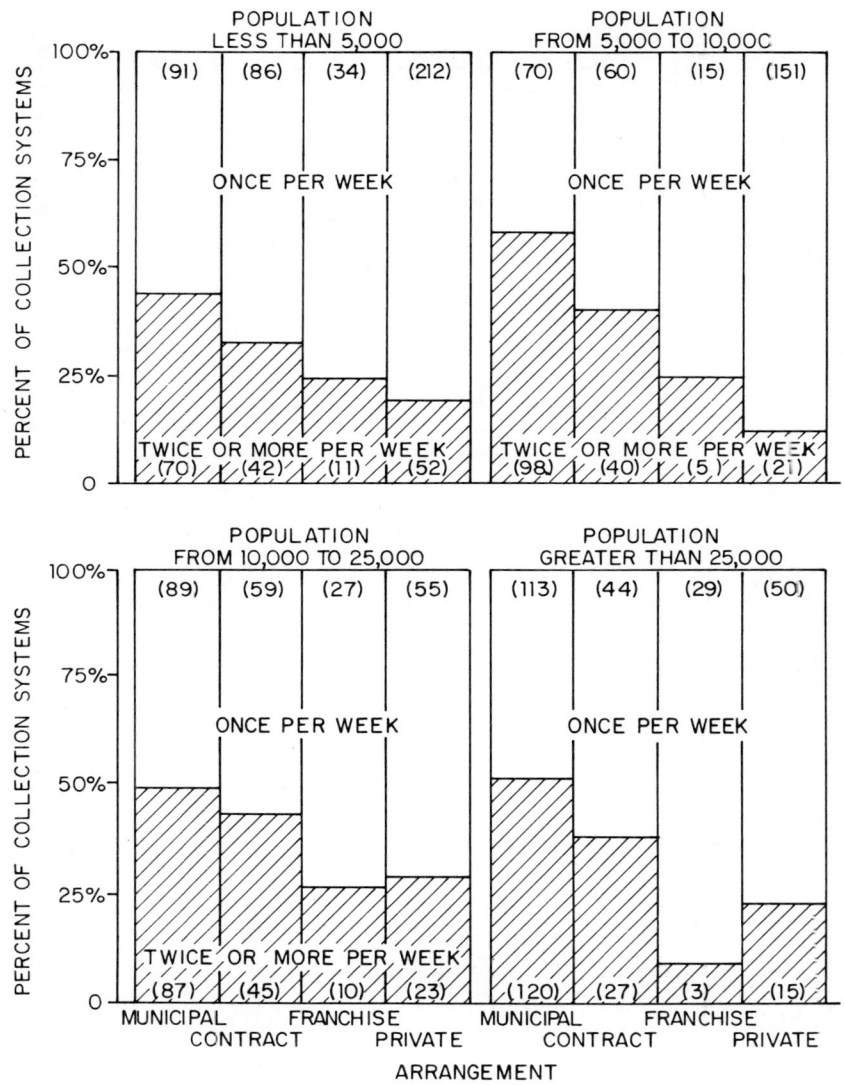

Figure 5-1. The Frequency of Collection by Arrangement, for Different Sizes (Numbers of Cities are in Parentheses)

willing to pay for, and this raises the question of the cost of service as a function of service level. This subject is discussed more extensively in Chapter 7, but it is worth noting that the higher cost of a higher level of service is generally more apparent when the private sector provides service (whether under a contract, franchise, or private arrangement). Supporting evidence for this assertion is

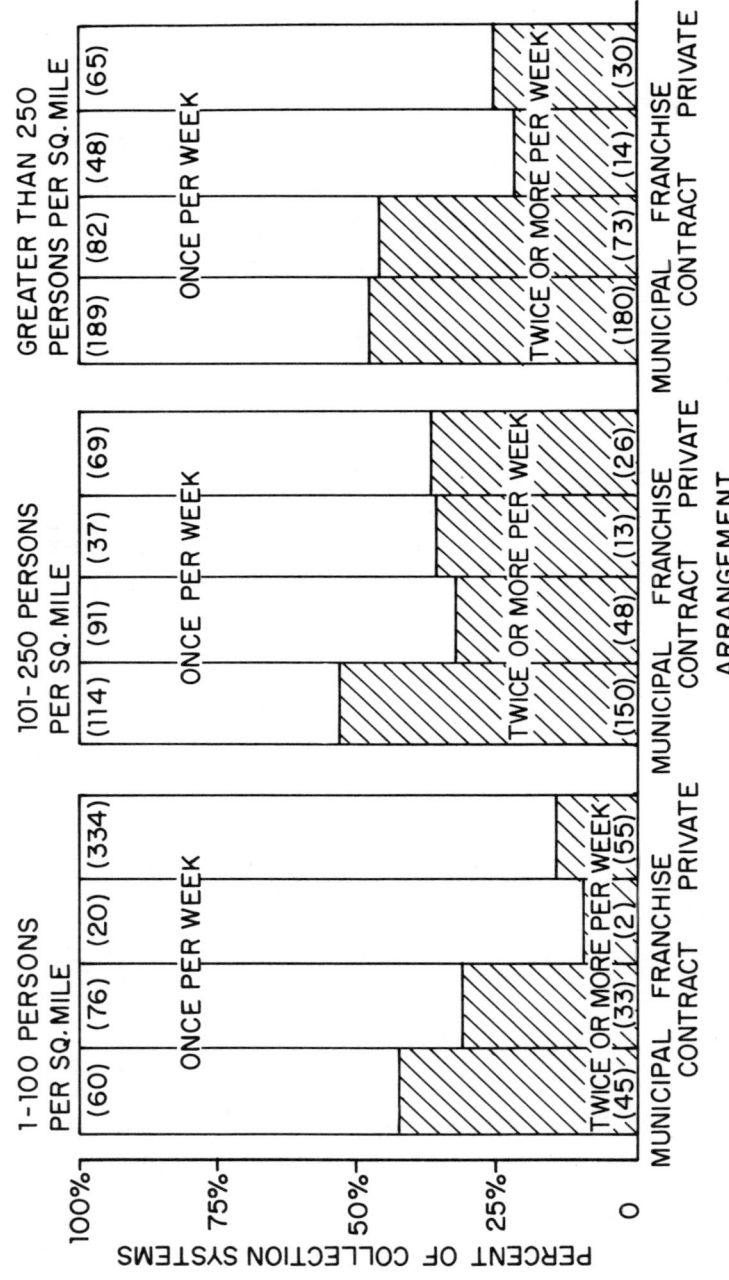

Figure 5-2. The Frequency of Collection by Arrangement, for Cities of Different Population Densities (Numbers of Cities are in Parentheses)

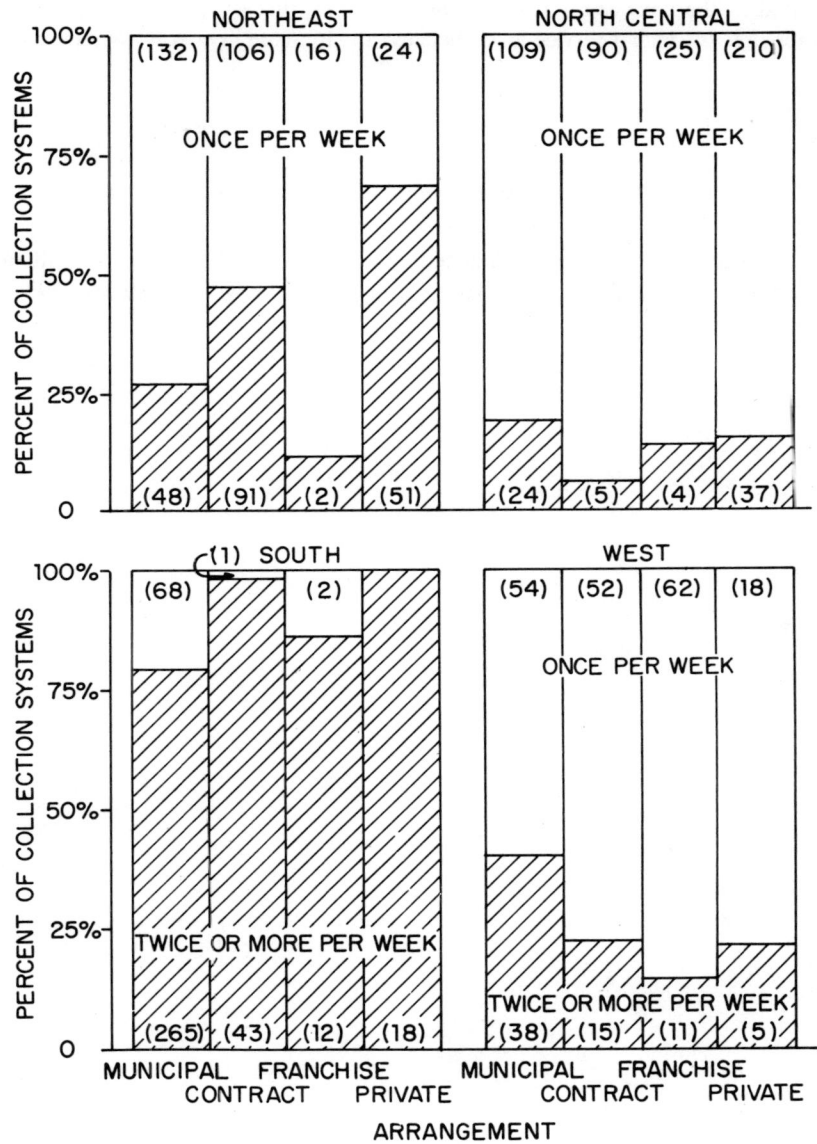

Figure 5-3. The Frequency of Collection by Arrangement, for Cities in Different Regions of the Country (Numbers of Cities are in Parentheses)

presented in the next chapter. It may be, therefore, that a municipality yields to the pressures for higher service levels because the increase in cost is not as visible.

Notes

1. William A. Niskanen, *Bureaucracy and Representative Government* (Chicago: Aldine, 1971).

2. George E. Peterson, "Finance," in *The Urban Predicament*, William Gorham and Nathan Glazer, ed. (Washington, D.C.: The Urban Institute, 1976).

6

Financing Solid Waste Collection

E.S. Savas, Daniel Baumol, and William A. Wells

This chapter examines the different sources of funds used to finance solid waste collection services. The principal modes of payment are the following:

1. Municipalities or other government units *tax* the user to pay for government-provided or contract services;
2. Municipalities and private firms charge the user a *flat fee* for the service regardless of service level;
3. Municipalities and private firms charge the user a *variable fee* for the service, the charge varying with either the quantity of refuse collected or the frequency or location of service provided. It should be noted that where there is a user charge levied by government, a tax-derived subsidy may also be used, either explicitly or implicitly.

The mode of payment may affect the cost of collection both directly and indirectly, and it may influence the behavior of the service recipient. The payment mode will determine whether the cost of billing is large or small, thereby affecting the cost of service directly, and it will also affect the cost directly to the extent that tax payments for refuse collection service are deductible for federal income tax purposes but user charges are not. The payment mode may affect the amount of refuse set out for collection, it may determine the demand for high-frequency service and for convenient service (for example, backyard vs. curbside collection), it may affect the extent to which the service recipient effectively monitors the performance of the service provider and causes the latter to maintain efficient and effective service, and it may influence the extent to which a service recipient will attempt to avoid the cost of regular refuse collection.

This chapter discusses four topics: the current prevalance of the different payment modes; the effects of payment mode on inducing individual households to purchase organized collection services; the effect of payment mode on the cost of refuse collection; and the relation between user charges and the cost of refuse collection.

Current Practices for Financing Solid Waste Collection

Financing methods fall into three categories: taxes and two types of user charges or fees, flat fees and variable fees. (Self-service, where the service recipient in effect "pays" for service by "donating" his labor, is excluded from this discussion.) The extent to which each of these payment modes is used is shown in Table 6-1. A user charge, whether flat or variable fee, is the most common means of financing the service. This is true for all three of the service recipients in Table 6-1. For example, taxes are used in 42.4 percent of cities to finance refuse collection from small residences, whereas user charges are imposed in 57.2 percent of cities.

With respect to flat and variable fees, the former are much more widely used than the latter for small-residential service, but the reverse is true for commercial collection. This is to be expected, as there is much variation in the amount of service needed by the commercial establishments in a community, in contrast to the relatively uniform service needs of small residences. Refuse collection from multiple dwellings is financed almost equally by taxes, flat fees, and variable fees.

Table 6-2 shows the fraction of people using each payment mode. It is derived from the same data as Table 6-1, but weights the cities according to their populations. More than half the people who live in residences other than multiple dwellings pay for their collection service through taxes. This is a consequence of the fact that large cities are more likely to have municipal collection financed by taxes.

This pattern is shown more clearly in Table 6-3. Generally, as city size increases, the fraction of cities that finances its services by taxes also increases. The use of flat fees charged by the municipality remains approximately constant for cities of all sizes. Similarly, the use of variable fees is constant regardless of

Table 6-1
Number and Percentage of Collection Systems Using Each Payment Mode, by Service Recipient

	Payment Mode				
Service Recipient	Tax	Flat Fee	Variable Fee	Don't Know	Total
Small Residences	42.4% (411)	43.8% (425)	13.4% (130)	0.4% (4)	100.0% (970)
Multiple Dwellings	34.0 (345)	31.0 (314)	33.7 (342)	1.3 (13)	100.0 (1014)
Commercial Establishments	31.8 (316)	16.3 (162)	51.8 (514)	0.1 (1)	100.0 (993)

Notes:
1. Number of cities shown in parentheses.
2. The local definition of multiple dwelling was used for each city. All other dwellings comprise the category "small residences."
3. Excludes self-service.

Table 6-2
Percentage of Residences Using Each Payment Mode

	Payment Mode				
Service Recipient	Tax	Flat Fee	Variable Fee	Don't Know	Total
Small Residences	52.6%	37.0%	10.1%	0.2%	99.9%
Multiple Dwellings	38.1	24.8	35.5	1.5	99.9

city size. These findings are noteworthy insofar as they suggest that large cities find it no more difficult or inconvenient than small cities to administer a system of user charges and variable service levels. (As city size increases, the fraction of cities paying for service by flat fees to private firms decreases, but this merely reflects the fact that municipal collection increases and collection by private providers declines as city size increases.)

Table 6-3 reveals that most cities with a mayor-council form of government rely on taxes to finance the service, while council-manager cities employ user charges to a much greater extent.

There are strong regional differences in the mode of payment. Northern municipalities in the East and Midwest rarely collect user charges from their residents while southern and western municipalities do so to a significant degree. The West is remarkable in that very few of its cities finance the service through taxes. These prominent differences suggest underlying philosophical differences among the regions; the South and the West appear to consider residential refuse collection as an individual good, to be paid for on an individual basis, while the North and East consider it more as a public good, to be financed collectively through taxes. This difference, in turn, may derive from the fact that in the South and West, with relatively low population densities, there are few external costs imposed on neighbors by poor refuse-collection practices, whereas in the more densely populated northeastern quadrant of the country there are substantial externalities and therefore collective action is required. This is discussed further in the next section.

It is particularly instructive to examine how the payment mode varies between and within arrangements. Table 6-4(a) shows that both municipal and contract collection of residential refuse (excluding multiple dwellings) are more often financed by taxes than by user charges, although the latter are quite common. More than two-thirds (67.4 percent) of contract arrangements are financed in this way, and 58.2 percent of municipal arrangements. The fact that user charges are even more common for contract collection than for municipal collection suggests that cities find it easier or more desirable to pass along the cost of contract collection—rather than municipal collection—to residents as a direct charge. This is explored further in a later section.

Table 6-3
Mode of Payment for Refuse Collection from Other-than-Multiple Dwellings

	Total		Tax		Municipal Flat Fee		Private Flat Fee		Municipal Variable Fee		Private Variable Fee	
	No.	Percentage	No.	Percentage	No.	Percentage	No.	Percentage	No.	Percentage	No.	Percentage
Total	2069	100.0	747	36.1	410	19.8	648	31.3	53	2.6	211	10.2
Population Group												
>500,000	14	99.9	10	71.4	1	7.1	3	21.4	0		0	
250,000-499,999	24	100.0	11	45.8	6	25.0	1	4.2	3	12.5	3	12.5
100,000-249,999	88	100.1	51	58.0	19	21.6	8	9.1	3	3.4	7	8.0
50,000-99,999	150	100.0	69	46.0	34	22.7	26	17.3	7	4.7	14	9.3
25,000-49,999	169	100.1	64	37.9	35	20.7	46	27.2	6	3.6	18	10.7
10,000-24,999	429	100.0	180	42.0	96	22.4	97	22.6	10	2.3	46	10.7
5,000-9,999	506	100.1	173	34.2	101	20.0	172	34.0	11	2.2	49	9.7
2,500-4,999	689	99.9	189	27.4	118	17.1	295	42.8	13	1.9	74	10.7
Geographic Region												
Northeast	763	100.0	356	46.7	43	5.6	287	37.6	2	0.3	75	9.8
North Central	577	100.0	203	35.2	38	6.6	257	44.5	6	1.0	73	12.7
South	438	99.9	157	35.8	224	51.1	30	6.8	10	2.3	17	3.9
West	291	100.0	31	10.7	105	36.1	74	25.4	35	12.0	46	15.8
Metro/City Type												
Central	279	100.0	150	53.8	64	22.9	32	11.5	10	3.6	23	8.2
Suburban	1790	100.0	597	33.4	346	19.3	616	34.4	43	2.4	188	10.5
Form of Government												
Mayor-Council	798	100.1	431	54.0	137	17.2	157	19.7	18	2.3	55	6.9
Council-Manager	615	100.0	177	28.8	232	37.7	116	18.9	32	5.2	58	9.3
Commission	58	99.9	26	44.8	22	37.9	1	1.7	3	5.2	6	10.3
Town Meeting	65	100.0	30	46.2	2	3.1	27	41.5	0		6	9.2
Rep. Town Meeting	14	100.1	6	42.9	0		4	26.6	0		4	26.6
Don't Know	519	100.0	77	14.8	17	3.3	343	66.1	0		82	15.8

Table 6-4
Payment Mode by Arrangement, for Residential Refuse Collection

A. For Small Residences

Payment Mode	Arrangement											
	Municipal			Contract			Franchise			Private		
	No. of Cities	Percentage of Row	Percentage of Column	No. of Cities	Percentage of Row	Percentage of Column	No. of Cities	Percentage of Row	Percentage of Column	No. of Cities	Percentage of Row	Percentage of Column
Tax	444	60.5	58.2	290	39.5	67.4	100	NA	NA			
Flat Fee	291	27.6	38.1	115	10.9	26.7	51	9.5	66.2	547	51.9	77.4
Variable Fee	28	10.6	3.7	25	9.5	5.8	51	19.3	33.8	160	60.6	22.6
Total	763	37.2	100.0	430	21.0	99.9	151	7.4	100.0	707	34.5	100.0

	No. of Cities	Total Percentage of Row	Percentage of Column
Tax	734	100.0	35.8
Flat Fee	1053	99.9	51.3
Variable Fee	264	100.0	12.9
Total	2051	100.1	100.0

B. For Multiple Dwellings

Payment Mode	Municipal			Contract			Franchise			Private		
	No. of Cities	Percentage of Row	Percentage of Column	No. of Cities	Percentage of Row	Percentage of Column	No. of Cities	Percentage of Row	Percentage of Column	No. of Cities	Percentage of Row	Percentage of Column
Tax	345	68.0	54.2	162	32.0	64.3		NA	NA			
Flat Fee	181	28.2	28.4	62	9.6	24.6	63	9.8	49.2	336	52.3	46.7
Variable Fee	111	18.9	17.4	28	4.8	11.1	65	11.1	50.8	383	65.2	53.3
Total	637	36.7	100.0	252	14.5	100.0	128	7.4	100.0	719	41.4	100.0

	No. of Cities	Total Percentage of Row	Percentage of Column
Tax	507	100.0	29.2
Flat Fee	642	99.9	37.0
Variable Fee	587	100.0	33.8
Total	1736	100.0	100.0

Variable fees are relatively rare in municipal and contract collection, but quite commonplace in franchise and private collection although they are, even there, a minority source of financing. A third of franchise arrangements and almost a quarter of private arrangements involve variable fees, implying that residents have some degree of choice as to service level under those arrangements, more so than under municipal and contract collection.

Payment modes for servicing multiple dwellings (see Table 6-4[B]) follow a pattern that is generally similar to that for financing service to small residences.

Mandatory Refuse Collection

Frequently it is in the interest of the community at large to require that its residents have regular refuse collection service. All citizens will then gain the aesthetic and health benefits associated with only small and short-term accumulations of refuse. Usually benefits from universal collection are greatest in large, densely populated cities, for in these cities diseases are easily communicable, and anyone refusing regular collection services has more neighbors to offend. Smaller and more rural cities without these pressing needs may not even want universal organized collection because they may wish to provide their residents with the option of saving money by providing their own service.

Those who do not pay for and receive organized service may dispose of their refuse themselves in a proper manner. However, some people dispose of it by illegal dumping or littering, or place it in refuse containers belonging to others, thereby imposing the costs (financial and other) of refuse accumulation or disposal on their neighbors. Where the problem is severe, such people must be coerced into having regular collection. This coercion can take place either by financing collection in a way that makes evasion of payment difficult and avoidance of service pointless, or by mandating regular refuse collection by an approved collector.

Of the payment modes in use, taxation is the hardest for a resident to avoid. Tax evasion is a serious offense and, because the tax used is generally a property tax, it is easy for government to collect this tax by threat of foreclosure.

Evading payment of municipal bills is also difficult. Not only does refusal to pay force a resident to confront the government directly, but in cases where joint billing is used to finance, for example, both water systems and refuse collection, the householder runs the risk of losing his water supply along with his refuse collection service. Both municipal and contract collection readily lend themselves to mandatory service.

At the other end of the spectrum from taxes is billing by a private firm to finance refuse collection. This method provides a relatively easier means for households to escape payment for refuse collection. Delinquent householders are in the position of defying a private company rather than a governmental entity.

Furthermore, the company has little recourse except a slow and expensive judicial process and the termination of collection service. In cases where households have a choice of companies, even when they are required to pay for organized collection, it is administratively difficult to attempt to examine customer lists in order to identify and apprehend service evaders.

For these reasons, mandatory collection is more difficult to effect by means of franchise or private collection than by municipal or contract collection. Nevertheless, some communities with franchise collection seem to be quite successful in enforcing mandatory service by working closely with the private firm to assure collection of outstanding bills, to coerce nonsubscribers identified by private firms, and to track down the people responsible for illegal dumping.

The Effect of Payment Mode on the Cost of Refuse Collection

The mode of payment affects the cost of collection both directly and indirectly. Indirect effects are those that alter the behavior of the customer and thereby change the refuse collection practices and the resultant costs. Direct effects are those that have a clear and direct impact on the cost of collection.

Indirect Effect on Consumer Behavior

Some economists have argued that when service consumers must pay a fee which increases as the quantity of service consumed increases, they will demand less service than other consumers. In terms of residential solid waste collection, this means that one could expect households to generate less refuse, have less frequent collection, and be more likely to have curbside collection if they pay for refuse collection with a variable rather than a flat fee or taxes.

This hypothesis is not supported by empirical data. There is little relationship between the mode of payment and waste generation (Table 6-5), the frequency of collection (Table 6-6), or the pickup location (Table 6-7).

Table 6-5
Relationship between Mode of Payment and Amount of Refuse Generated

Refuse Generation (in tons per household per year)	Payment Mode (Number and Percentage of Cities)					
	Tax		Flat Fee		Variable Fee	
	No.	Percentage	No.	Percentage	No.	Percentage
Less than 1.5	40	60.6	52	48.6	15	55.6
1.5 and more	26	39.4	55	51.4	12	44.4
Total	66	100.0	107	100.0	27	100.0

Table 6-6
Relationship between Mode of Payment and Frequency of Collection

Frequency of Service	Payment Mode (Number and Percentage of Cities)					
	Tax		Flat Fee		Variable Fee	
	No.	Percentage	No.	Percentage	No.	Percentage
More than twice per week	66	8.8	88	8.4	89	34.1
Twice per week	256	34.3	343	32.6	30	11.5
Once per week	424	56.8	622	59.1	142	54.4
Total	746	99.9	1053	100.1	261	100.0

Table 6-5 shows that those who pay variable fees are about as likely to generate (relatively) little refuse as are those who pay by tax or flat fee. Table 6-6 indicates that payment of a variable fee is not associated with low-frequency service. Finally, Table 6-7 reveals that the low-cost collection option, curbside or alley service, is not associated with a variable fee structure. (In fact, those who pay a variable fee are more likely to have a convenient [to themselves] pickup location than those who pay taxes or flat fees. This can be explained by reference to family income. In Chapter 5 it was shown that higher-income families are more likely to have backyard or frontyard pickup than lower-income families. It seems logical that high-income families are better able to demand and pay for specialized service instead of uniform service.)

Direct Effects of Payment Modes on Cost

The mode of payment has a direct effect on the cost to the household in two ways: it affects the cost of billing, and it affects the federal income tax status of the payment. These two effects are first discussed separately and then their combined impact on cost in a typical city is discussed.

Administrative Billing Expenses. The three major billing alternatives used to charge for solid waste collection are: indirect billing through general tax revenues such as income and property taxes; combination billing with other services such as water, electricity, or sewers; and separate billing.

Of these three options, indirect billing or tax collection is by far the cheapest. This is the case simply because taxes are collected whether or not they finance refuse collection; the marginal cost of billing for refuse collection is effectively zero in this case.

Combination billing is an option which is generally considered to be

Table 6-7
Relationship between Mode of Payment and Pickup Location

Location of Pickup	Payment Mode (Number and Percentage of Cities)					
	Tax		Flat Fee		Variable Fee	
	No.	Percentage	No.	Percentage	No.	Percentage
Curbside or alley	478	64.2	555	52.6	65	24:9
Other	266	35.8	501	47.4	196	75.1
Total	744	100.0	1056	100.0	261	100.0

available only to governmental units (although in principle, government could pay a private electric, gas, water, or telephone utility to collect sewer or refuse collection charges). Such billing is less expensive than direct billing for two reasons. First, when a single bill is used for several services such as water and electric power, the cost of billing is divided among the several services. Further, when one bill covers more than one service, the delinquent customer runs the risk of losing several services; someone willing to forgo refuse collection may not be willing to surrender his water supply.

Billing costs were studied in thirty-nine cities that had municipal collection, paid for by user charges. It was found that billing costs are generally less than 5 percent of total refuse collection costs, with a mean of 3.1 percent. Many of these cities used combination billing. Figure 6-1 shows the ratio of billing costs to total costs for the cities.

Separate billing is usually the most expensive procedure. The entire cost of billing is attributable directly to refuse collection, and since the chance of losing collection service by failing to pay is not as ominous as the risk of either evading tax payments or going without necessities such as water or electricity, more bad debts can be expected.

Franchise and private collection, as defined here, must generally rely on separate billing, the most expensive alternative. Several private firms have reported that their billing expenses amount to 10 to 15 percent of the cost of refuse collection. Generally, only municipal and contract collection arrangements are able to take advantage of the low-cost alternatives, using taxes or municipally imposed user charges that are billed jointly with other services. Nevertheless, in at least one instance a private firm adopted a particularly creative approach to avail itself and its customers of the economies of joint billing. Operating under a franchise arrangement, the firm entered into a contract with the city and paid the latter to handle its billing work, although the city was not responsible for bad debts and did not use its police powers to enforce payment. In another city with mandatory franchise collection, if a bill remains unpaid despite a series of actions, the private firm may ask the city to intervene, and, after investigating the firm's complaint, the city will add the

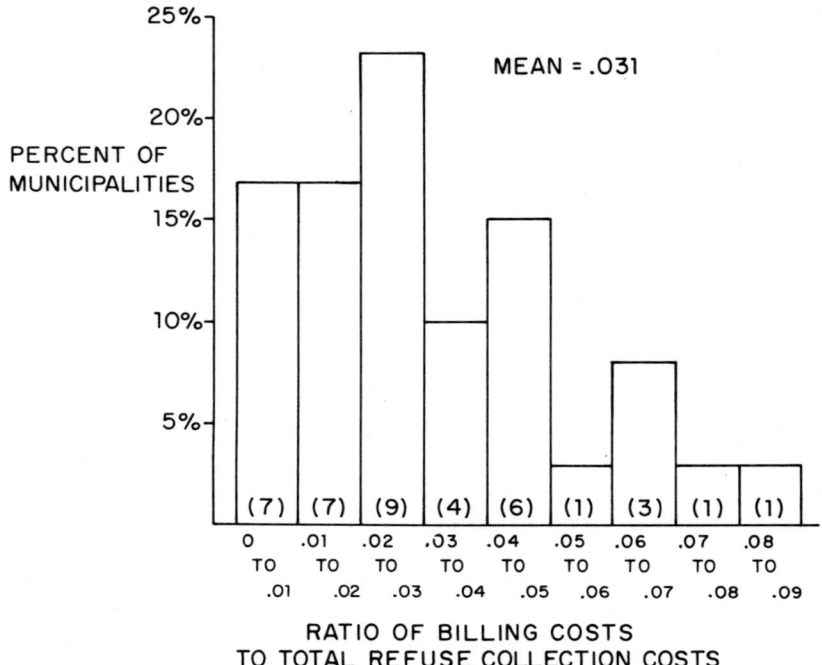

Figure 6-1. Distribution of the Ratio of Billing Costs to Total Cost, for 39 Municipal Solid Waste Collection Systems (Numbers of Cities are in Parentheses)

delinquent charge to the tax bill for the property, thus requiring payment. Innovative financing procedures such as these are worthy of consideration in all cities where payment is by separate billing. After all, whatever the cost of billing, the service recipient ultimately pays for it.

The Effect of Federal Income Tax Policy on the Cost of Collection. The federal government permits taxpayers to deduct the cost of local taxes from reported income for income tax purposes. This means that when collection is financed with tax revenues, this cost is tax deductible. On the other hand, user charges are not deductible. Therefore, if a family pays $50 for refuse collection through taxation and it has a marginal tax rate of 25 percent, the "aftertax" cost to that household is only $37.50. If the $50 is paid in the form of a user fee, the family will get no reduction in its income tax payment and will pay the full cost of refuse collection.

Two facts about this tax deduction policy are immediately apparent. First, it is clear that this deduction is worthless for the family that does not itemize income tax deductions. Second, for those families that do itemize deductions (usually high-income families) the value of these deductions depends on the family's marginal tax rate; a $50 deduction is of more value to a family that pays a 60 percent rather than a 30 percent tax rate on its last dollar earned.

Analysis of an Average Community. In this section, three possible payment modes—tax collection, billing by the city, billing by a private firm—are examined hypothetically in a typical community for the purpose of calculating the direct effect of payment mode on cost. Cost is measured in aftertax dollars in each case. It is assumed that for each payment mode the price of collection (excluding the cost of billing) is $34 per household per year.

In the case of municipal billing, there will be a billing expense of 3 percent (see Figure 6-1) so that total collection costs are $35.02 (1.03 × 34.00). In the case of private billing, the billing expense is assumed to be 10 percent so that the total cost of collection is $37.40 (1.10 × 34.00). Because user fees are not allowable deductions on the income tax, these are the aftertax costs of refuse collection in these two cases.

Calculating the aftertax cost of collection when the service is financed with property tax revenues is more complicated. Before a methodology is even discussed, two assumptions must be made: first, it is assumed that the assessed value of property holdings is directly proportional to income (see Netzer[1]); second, it is assumed that the community has an income distribution similar to that of the U.S. urban population in 1970. (This distribution is shown in Column 1 of Table 6-8.)

The first step in the process of determining the effect of tax deductibility on costs is to calculate the total beforetax cost of collection for households in each income bracket. Since property taxes are assumed to be proportional to income, if any one group receives one-half of the community's income, it will pay one-half of the community's waste collection bill. The result of this calculation appears in Column 3 of Table 6-8.

The remaining step in calculating the aftertax cost of collection in this community is to find out how much of the collection cost in Column 3 is reflected in reduced income tax bills. To estimate these figures, it is essential first to know what percentage of people take advantage of the deductibility of these charges by itemizing their federal tax deductions (shown in Table 6-8, Column 4). Then it is essential to know the marginal tax bracket of people in each income bracket. These are shown in Column 5. The average tax savings per household and the net cost of refuse collection, after taxes, can then be calculated, as shown in the last two columns of Table 6-8.

Now it is possible to summarize and compare the costs of collection in a community where all factors remain constant except the means used to finance refuse collection. When private billing is used, the aftertax cost of collection is $37.40 for all households. When municipal billing is used, the cost is $35.02.

Table 6-8
Estimation of Aftertax Cost of a Refuse Collection Service That Is Financed by Property Taxes

Income Bracket	(1) Percentage of Population[a]	(2) Average Income Times Percentage of Population[b]	(3) Per Household Cost of Collection Before Taxes[c]	(4) Proportion of Persons Itemizing Tax Deductions[a]	(5) Marginal Tax Bracket (Proportion of Income)[d]	(6) (3) × (4) × (5) Average Federal Tax Savings Per Household	(7) (3) − (6) Per Household Cost of Collection After Taxes
Under $2,000	15.0%	15,000	$ 3.30	.012	.15	$ .01	$ 3.29
$2,000-3,999	12.9	38,700	9.89	.066	.16	.10	9.97
$4,000-7,999	23.7	142,200	19.78	.214	.19	.80	18.98
$8,000-11,999	19.5	195,000	32.97	.444	.22	3.22	28.75
$12,000-14,999	10.7	144,450	44.52	.549	.25	6.11	38.41
$15,000-19,999	10.1	176,750	57.70	.710	.28	11.47	46.23
$20,000-49,999	07.4	259,000	115.41	.869	.37	37.10	78.31
$50,000 and over	00.8	60,000	247.31	.967	.55	131.53	115.78
Total (or Avg.)	100.1	1,031,100				6.44	27.56

[a]U.S. Department of Treasury, Internal Revenue Service, *Statistics of Income 1972: Individual Income Tax Returns*, publication 79 (1-75), pp. 94, 121. It is assumed that each tax return represents a household.

[b]Average income is assumed to be the midpoint of the range and for the over $50,000 bracket to be $75,000.

[c]For example, for the $2,000-3,999 income bracket, the formula used is $(34.00 \times \frac{38,700}{1,031,100} \times \frac{1}{12.9}) = 9.89$

[d]U.S. Department of the Treasury, Internal Revenue Service, 1975 Instructions for Form 1040, p. 25, Schedule Y for Married Taxpayers Filing Joint Returns.

When property tax revenues are used, the cost of collection varies by income bracket, yet, as is shown in Figure 6-2, the cost of collection is less expensive than other options for all families whose taxable income is less than $13,000.

As Table 6-9 shows, a substantial fraction of the total cost of this hypothetical community's refuse collection service can be saved by financing the service through taxes (assuming that the marginal cost of tax collection is zero). Municipal billing results in a cost 27 percent higher, and private billing 36 percent higher, than the aftertax average cost of tax-financed service.

While this discussion shows that taxation is the least costly payment mode for the community as a whole, and is particularly economical for households with taxable incomes less than $13,000 per year, the latter are, in effect, being subsidized by households with taxable incomes higher than that. This may be considered equitable or not, depending on one's values. Picking up a can of refuse from the curb costs the same whether the contents belonged to a rich or a poor family, and most cities (as shown in Table 6-1) appear to operate on the equitable principle of equal payment for equal service (that is, a user charge). An extreme case illustrates the inequity of paying for services by property tax: an elderly widow who lives alone in a large mansion and puts out a tiny bag of refuse every week would pay much more for refuse collection than a robust family of twelve that lives in a tumbledown shack and generates large volumes of garbage and trash every day.

For a community which has an income distribution very different from that in the hypothetical community, paying for service by taxes may offer advantages to all residents. This is clearly the case in a community of upper-income families, where a tax-financed service would be subsidized, in effect, by taxpayers elsewhere because of the tax reduction that local residents would obtain. Even so, although this is a clear and compelling argument in principle for tax-financed service in such a community, practically speaking, the actual dollar benefit for such families may be too small to outweigh other possible reasons for choosing a system that requires a user charge.

User Charges and the Cost of Collection

What is the relationship between a user charge and the cost of service? If a city finances the service through a user charge, whether it has municipal collection or contract collection, does the city "make money" on this business, by charging more than it has to pay, does it tend to subsidize the service (through taxes) by charging less than the full cost, or does it try to break even, by charging its residents the full cost, no more and no less?

These questions were examined for cities that levy user charges, billing their residents for municipal or contract service.

In numerous cases, the prices that municipalities charge residents for refuse

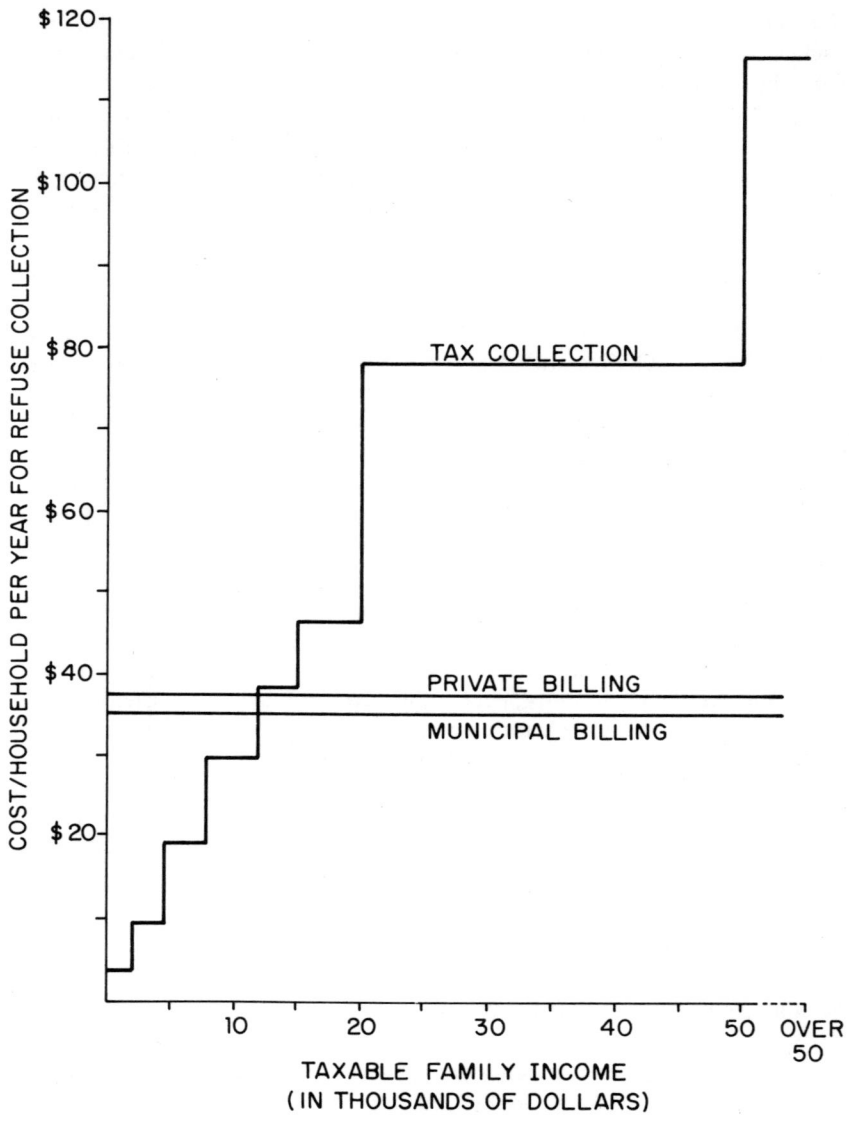

Figure 6-2. Total Cost of Refuse Collection Varies by Mode of Payment and Family Income because of Deductibility of Property Taxes on Federal Income Tax Returns

Table 6-9
Summary of Effect of Payment Mode on the Aftertax Cost per Household for Refuse Collection

Mode of Payment	Net Cost per Household after Taxes	Net Cost Relative to Tax-Financed Service
Private Billing	$37.40	136
Municipal Billing	35.02	127
Tax Collection	27.56	100

collection bear little relationship to the actual cost of providing this service. In one large city, for example, the city charges residents $48 per household per year for collection service while the city pays a firm only $29.50 per household per year to provide this service. At the other extreme of the spectrum, another city has a municipal collection system that incurs costs of $50 per household per year, yet residents are charged only $12 per year for this service. (See Chapter 7 for a discussion of how service costs were determined.)

Thirty-seven cities in the sample had municipal collection and levied a user charge. In these cities the average ratio of collection cost to user charge was 1.26 (standard deviation: 0.76); that is, the actual cost of service was substantially greater than the price charged by the city to its residents. The comparable figure for the thirteen cities in the sample with contract collection (and a user charge collected by the city) is 0.96 (standard deviation: 0.23); that is, the cost of collection was slightly lower than the user fee.

One would expect this ratio (of costs to user charge) to be slightly less than one because the cost, as calculated, excluded the cost of disposal, and a user charge for refuse collection service, in principle, is implicitly intended to pay both for collection and disposal. Therefore, it appears that cities with user charges and contract collection explicitly pass on the full costs of service to their residents, but cities with user charges and municipal collection often do not impose fees large enough to cover the cost of service and have to make up the difference with taxes or other revenues.

The situation is revealed more clearly in Figure 6-3. There is considerable variation in the ratio of costs to fees, and little relation between the two. (If there were a relation, one would expect the data to fall along the 45° line.) One explanation for the variation is that some cities deliberately treat refuse collection as an income-generating service, while others are motivated to subsidize the service for some residents by relying partly on taxes and partly on user charges. However, this explanation fails to account for the much lower variation found among cities with contract collection, as shown in Figure 6-4.

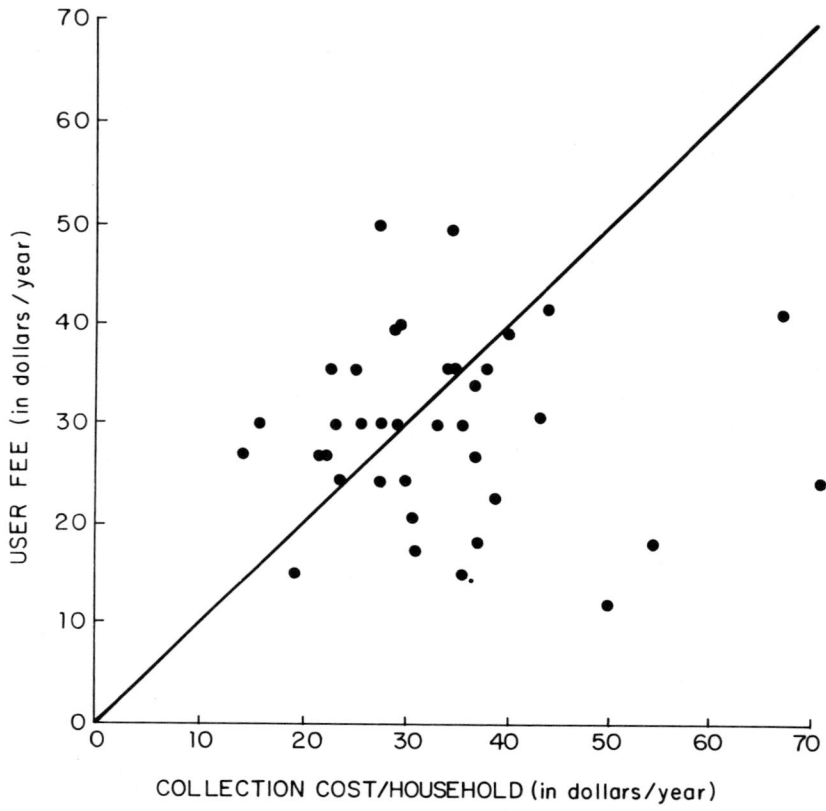

Figure 6-3. User Fees vs. Collection Costs for Cities with Municipal Collection

One is tempted to explain this disparity in the following way: *Cities with contract collection know how much the service costs; cities with municipal collection do not.* Under contract collection the cost of service is obvious, being stated in a public document. However, cities with municipal collection have cost-accounting practices that fail to indicate the full cost of service, with the result that their actual costs are considerably in excess of the costs indicated in their budgets.[2] (This issue is examined in a later section.)

Therefore, the simplest explanation of Figures 6-3 and 6-4 is that cities with contract collection generally set their fees equal to their costs, but cities with municipal collection, not knowing their costs accurately, are unable to set their fees accordingly.

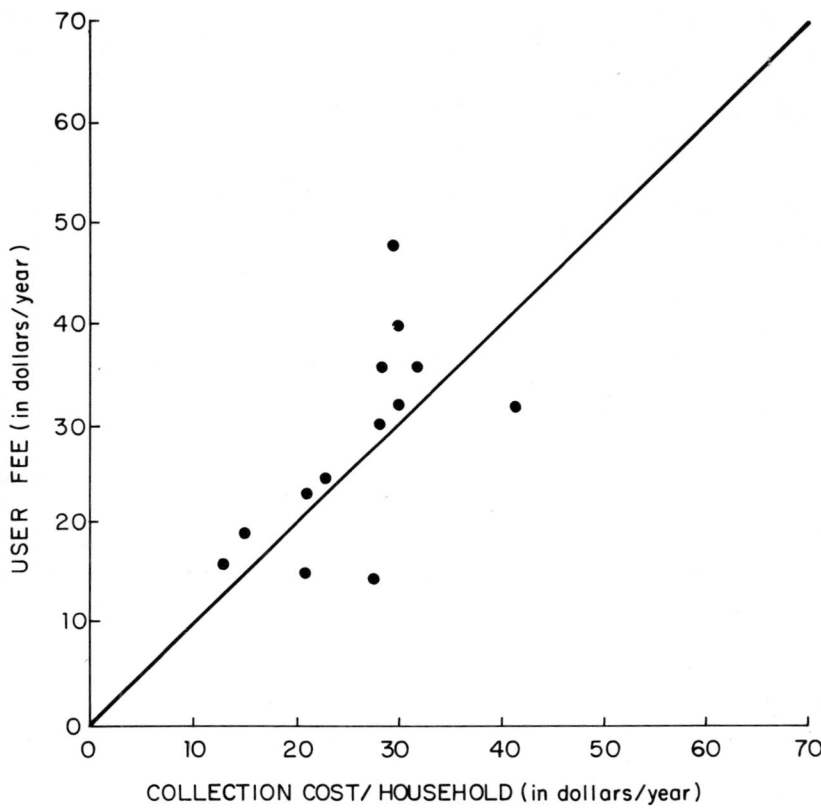

Figure 6-4. User Fees vs. Collection Costs for Cities with Contract Collection

Summary

This chapter has examined the prevalence and effects of taxes and user charges—both flat and variable fees, levied by governments or by private firms—for financing solid waste collection. It was found that user charges are more common than taxes for both residential and commercial collection. However, contract collection and municipal collection of residential refuse are more often paid for by taxes than by user charges. User charges levied by local governments are employed to a similar extent regardless of city size. Council-manager cities utilize fees much more than cities with a mayor-council form of government. There are large differences among regions in their payment modes.

Because of external costs and benefits associated with the proper removal of solid waste, many cities require their residents to receive organized service. Tax collection and, to a lesser extent, municipal billing permit relatively effective enforcement of this policy.

The use of a fee that varies with the level of service does not appear to have a significant impact on the amount of refuse offered for collection, or the choice of service level.

While taxes used to pay for refuse services can generally be collected at no extra cost to the community, user charges require a billing procedure that adds to the full cost of service; about 3 percent for municipal billing (which may involve joint billing for several services) and no less than 10 percent for billing by a private firm.

Federal income tax laws allow taxes, but not user charges, as deductions. This creates a considerable incentive for a community, particularly a wealthy one, to finance its service through taxes; other payment modes will impose costs that are 27 percent to 36 percent greater.

It was found that the user charge is closely related to the cost of service in cities with contract collection, but there was no such correspondence in cities with municipal collection. This is attributed to cities knowing how much contract collection costs, but not knowing very well how much municipal collection costs them.

Notes

1. Richard Netzer, *Economics of the Property Tax* (Washington, D.C.: Brookings Institution, 1966), chap. 3.

2. E.S. Savas et al., *Refuse Collection: Department of Sanitation vs. Private Carting* (New York: Office of the Mayor, City of New York, 1970).

7

The Cost of Residential Refuse Collection

Barbara J. Stevens

Introduction

Residential refuse collection in the United States accounts for a large portion of total expenditures for solid waste management. In 1970-1971, the National Commission on Productivity estimated that total expenditures for all refuse collection and disposal operations were approximately $3.5 billion; of the 227,000 persons employed by the industry, 100,000 were involved in the collecting of residential solid waste.[1] Using the ratio of employees in residential solid waste collection to total industry employees, one can estimate that perhaps 44 percent of the total $3.5 billion in solid waste expenditures, or $1.56 billion, was attributable to the collection and disposal of residential solid wastes. A 1975 study of refuse collection costs in 315 cities in the United States, with populations totaling 15,568,000, found that the average per capita cost for collection alone was $10.28. Projecting this number for the nation as a whole gives an estimate of the current costs of residential refuse collection—over $2 billion. For 238 of these 315 cities, data on the costs of refuse disposal were also obtained; the average annual per capita cost of disposal was $1.10. Thus, the total costs of refuse collection and disposal were about $11.40 per person per year in 1975, or approximately $2.5 billion for the country as a whole.

The preceding estimates of the costs of collection and disposal of residential refuse are clearly imprecise. The very empiricism of these estimates highlights one of the major problems in managing solid waste collection operations: the lack of regularly collected reliable data. The lack of reliable guidelines for solid waste management is especially critical, as even the rough estimates presented indicate clearly that the cost of refuse collection is rising—the four-year increase from $1.54 billion to $2.5 billion would be consistent with an annually compounded rate of cost increase of about 13 percent. If this rate of increase were to continue, the total costs of refuse collection would double every six years.

The importance of understanding the determinants of the cost of refuse collection and disposal is obvious to anyone interested in ensuring the delivery of such services at an ecologically safe and economically prudent level. The cost of refuse collection and disposal depends upon many factors, some of which are under the direct control of the manager of the refuse collection and disposal operation and some of which are not. For full understanding of the determinants of the cost of refuse collection it is necessary to know which individuals or collection bodies control which elements of cost. While it seems clear that cost minimization subject to environmental and aesthetic constraints is a desirable goal for the solid waste manager, only through understanding the abilities of various actors to influence costs can any facility in controlling costs be achieved.

To gain the necessary understanding of the relationships between the various components of the cost of refuse collection it is necessary to enumerate the components, to identify the agents with control over each, and to analyze the interrelationship between changes in individual components and total costs. The next section presents data on the cost of residential collection, disaggregated by five major components of refuse collection costs. These data were obtained using a consistent set of data collection instruments (see Appendix B) and trained interviewers. The data are presented in aggregated form as well as disaggregated by region and market size. The determinants of the costs of residential refuse collection are also discussed. The factors outside the control of the refuse collection manager—weather, topography, prevailing wage rates, and so on—are distinguished from those factors within his purview—technology, labor management, and so forth. Data are presented to highlight the relationship between these determining factors and the costs of refuse collection. The final section contains conclusions regarding the factors that have the most significant impact on refuse collection costs.

Cost Components in Residential Refuse Collection

Major Cost Components

One of the most common difficulties encountered in previous attempts to estimate the cost of refuse collection is the lack of reliable data. Ralph Stone summarized the problem neatly: "Reported cost data indicated wide variations among cities in both cost efficiency and accurate accounting. In fact, one of the most interesting conclusions to be drawn from an analysis of the same data is that many American cities have no way of accurately determining the productivity of their collection dollar simply because they fail to record adequate refuse quantity and manpower data."[2] Although the problem was recognized in 1969, it persists today. To a large extent the problem is not so much that a particular city does not know its own costs of refuse collection but rather that it finds it difficult to determine, in comparison with similar cities, whether its costs are below, above, or equal to the average.

Today many cities still do not know whether their true costs of refuse collection are relatively high or relatively low because of the wide differences in accounting procedures used by city governments, which often make it difficult for a city to determine its collection costs. For example, some cities lump all vehicle operation and maintenance expenses into a single category, while others allocate such expenses to the departments actually using the vehicles. Comparison of budget line items for refuse collection between such cities would clearly tend to bias conclusions regarding cost efficiency: those cities whose accounting procedures allocate few expenses to refuse collection would tend to appear, perhaps falsely, more efficient than those cities whose accounting procedures allocate all expenses associated with refuse collection directly to that department. Therefore, to provide useful cost information, the problem of differing accounting systems must be faced. This problem is most difficult when a city agency collects the refuse and less so when a private firm services the residents.

For cities serviced by a private firm, the fee charged by the firm (or the amount paid the firm by the city), after subtracting any expenses associated with disposal, may properly be considered the cost of refuse collection *to the residents served*. Note that this figure includes profits earned and taxes paid by the firm. For communities served by a public agency, however, the budget line item for refuse collection cannot properly be considered the true cost of refuse collection. Rather, on-site data collection, using a consistent method in all cases, is necessary to determine this cost accurately.

For cities serviced by private firms, costs were obtained, as appropriate, from city officials, private households, and the private firms, using a combination of telephone and mail surveys.[3] For both the public and private sectors, data were obtained using consistent definitions so that intercity variations in the cost of refuse collection could validly be made.

It should be stressed that in all cases the cost figures presented here are for collection only. For public service delivery, that includes all costs incurred to collect refuse and remove it to a disposal site or transfer station but excludes any costs of operating such disposal facilities. For the private service deliverer the price charged the household is adjusted downward by subtracting the disposal costs, if incurred by the private firm, in order to arrive at the cost for collection alone.[4] Such disposal costs might be based on loads disposed of, tons disposed of, or operating costs where the disposal facility is managed by the collector himself. In all cases, then, cost to the household for collection alone is less than (or equal to, in cases in which a private firm receives free use of a disposal facility) the price charged the household by the private firm.

For any collector of residential refuse, five basic categories of collection costs may be distinguished. These categories were selected for theoretical and practical reasons. For example, by separating labor expenses from equipment expenses, one may investigate the capital intensity of refuse collection technology; in addition, many cities have quite separate departments and/or accounting systems for labor and equipment expenditures so this separation is of practical significance as well.

Each of these cost components was identified from interviews and city records by a trained interviewer working on-site. First, labor costs were determined. These included direct labor costs, overtime, and that portion of the wages of supervisors, clerks, and other support workers which could properly be attributed to residential refuse collection. Second, total fringe benefit costs were obtained by summing the city's expenses for each available benefit. Third, expenses for vehicle maintenance and repair, office expenses, billing and insurance expenses, and any other miscellaneous operating expenses were collected and equated to operating costs. Fourth, a portion of the expenses associated with overhead agencies (such as the mayor, the finance department, payroll office, and so on) were allocated to refuse collection. Fifth, vehicle depreciation was calculated on a five-year, straight-line basis.

Description of the Data

Three different kinds of cities were included in the study: cities with municipal collection, where a public agency collected the refuse; cities with contract collection or mandatory franchise (MF) collection, where a private firm under contract or franchise with the city collected refuse from all residences; and cities with private or nonmandatory franchise (NMF) collection, where a single private firm collected refuse from some but not all of the residences (the remaining residents hauled their own refuse or received service from a competing firm).

Table 7-1 shows the composition of the sample of 315 cities. The cities receiving municipal collection are over twice as populous, on average, as the cities receiving service from private firms. The overall population of cities sampled, irrespective of service arrangement, ranged from 2,524 to 717,099.[5] Although the large cities sampled tended to employ municipal refuse collection, there is no practical or theoretical reason why large municipalities need employ this particular arrangement: for example, San Francisco and Boston use mandatory franchise and contract arrangements respectively, whereas Portland, Oregon, uses the private arrangement.

The prevalence of organizational arrangements in the cities sampled also varied by region of the country. Table 7-2 presents a correlation of organizational arrangement and region of the country. The numbers are the percentages of

Table 7-1
Sample of Cities Used for Analysis of Costs

Collection Arrangement	No. of Cities	Average Population	Population of Smallest City	Population of Largest City
Municipal	102	84,000	2,615	717,099
Contract and MF	92	39,000	2,524	530,831
Private and NMF	121	29,000	2,578	149,518
Total	315	50,000	2,524	717,099

Table 7-2
Percentage of Sample from United States Regions by Arrangement Type

	Organizational Arrangement			
Region	Municipal	Contract and MF	Private and NMF	Number of Observations
New England	35.7	32.1	32.1	28
Mid-Atlantic	23.4	29.7	46.9	64
East North Central	23.4	29.7	46.9	70
West North Central	45.5	0	54.5	11
South Atlantic	76.7	13.3	10.0	30
East South Central	77.8	22.2	0	9
West South Central	60.0	40.0	0	30
Mountain	21.7	34.8	43.5	23
Pacific	18.0	36.0	46.0	50
Total	32.4	29.2	38.4	315

See Appendix A for list of states in each region.

localities in each region with a given organizational arrangement; thus all percentages sum to 100 across each row. If there were no relationship between region and organizational arrangement, it would be expected that the percentages in each region's row would equal those of the entire sample. As is shown in Chapter 4, however, this is not the case. For example, whereas one would expect about a third of cities in each region to provide municipal service, in fact, such service is much more likely to be found in the South Atlantic (76.7 percent of localities in the sample from that region had municipal service), the East South Central (77.8 percent municipal), and the West South Central (60.0 percent municipal) regions.

Despite the differences in the distribution of arrangements by region, within each region each arrangement conducts its refuse collection operation in a roughly similar manner. For example, when prevailing wage rates are high, one would expect that refuse collection would tend to use a more capital intensive collection technology (larger vehicles, fewer men per vehicle, and so forth) than when wage rates are low. In general, such a relationship holds for each arrangement type in a region.

As Table 7-3 shows, the average crew size for the municipal arrangement is 3.2 men. The largest crews (3.7 men) are used in the South Atlantic and East South Central regions, where wages paid are below the average of $574. The smallest public crews are found in the Mountain and Pacific regions (2.2 and 2.1 men, respectively); as is expected, wages paid in these regions ($671 and $708) are well above the average of $574. For the contract (and mandatory franchise) arrangement, two of the three regions with the smallest crew sizes are associated

Table 7-3
Collection Technology, Wage Rate, and Arrangement

Region	Wage Rate			Crew Size			Truck Size (in cubic yards)			Percentage Loading in Rear		
	MU[a]	CO[b]	PR[c]	MU	CO	PR	MU	CO	PR	MU	CO	PR
New England	577 (10)	729 (9)	654 (9)	3.3	2.0	1.4	18.5	22.3	16.2	90	72	78
Mid-Atlantic	610 (15)	740 (19)	610 (30)	3.2	2.1	1.8	19.2	22.4	19.1	89	87	77
East North Central	735 (16)	682 (21)	714 (33)	3.4	2.1	1.6	18.9	22.4	19.6	100	82	86
West North Central	519 (5)	NA	657 (6)	2.7	NA[d]	1.7	16.9	NA	19.5	77	NA	83
South Atlantic	548 (23)	828 (4)	596 (3)	3.7	2.6	2.7	19.2	25.2	20.0	86	91	33
East South Central	482 (7)	514 (2)	NA	3.7	2.0	NA	16.6	19.0	NA	56	75	NA
West South Central	442 (18)	564 (12)	NA	3.0	2.4	NA	20.1	22.2	NA	83	63	NA
Mountain	671 (5)	507 (8)	737 (10)	2.2	2.4	1.7	25.0	23.1	22.2	37	87	76
Pacific	708 (9)	688 (18)	744 (23)	2.1	1.8	1.5	24.9	26.8	24.7	57	41	38
Overall	574	669	691	3.2	2.1	1.7	19.8	23.3	20.4	80	72	72

Number of cities in parentheses; applicable to all variables.
See Appendix A for list of states in each region.

[a]MU = Municipal
[b]CO = Contract and Mandatory Franchise
[c]PR = Private and Nonmandatory Franchise
[d]NA = No entries in cell

with wage rates above the average for that arrangement as a whole. Similarly, two of the three regions with the largest crew sizes had wages that averaged less than the overall average of $669. For the private and nonmandatory franchise arrangement, the two regions with the largest crew sizes also pay wages below the national average; 23 of 30 cities in the two regions with the smallest crews pay wages above the national average of $691 for the private arrangement. A similar pattern may be traced for truck capacity, with larger capacities per crewman associated with higher wages. The percentage of vehicles loading in the rear is well below the average for each arrangement in the Pacific region, where wages paid by each arrangement are above its respective national average. However, this relationship does not hold so clearly for the other regions.

Cost Components by Region, Market Size, and Arrangement

Actual data on the components of the costs of refuse collection were collected only for cities served by a municipal collection agency. However, the similarity of production technologies across organizational arrangements would indicate that the relative importance of each cost component may be broadly applicable to both public and private collection.

Table 7-4 presents the findings of the cost components for residential refuse collection. They are based only on data from the 101 cities with municipal collection.

The mean annual cost per household for this group of cities is $32.08. On average, 57 percent of the cost of refuse collection is attributable to labor costs, 12 percent to fringe benefit costs, 16 percent to operating expenses (of which 72 percent is attributable to the costs of vehicle operation), 9 percent to overhead costs, and 6 percent to depreciation of capital equipment. On average, fringe benefits amount to 21.9 percent of wages and salaries, although the figure goes as high as 56.4 percent in one instance. It is clear, then, that the cost of collection is highly sensitive to the quantity of labor employed and to its level of remuneration.

Wages paid to collectors are perhaps the simplest indicator of remuneration levels. This figure was obtained both for those cities served by public agencies and those served by private firms. The mean monthly wage, exclusive of fringe benefits, paid to a collector by public agencies is $574 and by private firms

Table 7-4
Per Household Costs of Residential Refuse Collection, in Cities with Municipal Collection in Dollars per Year

Cost Component	Mean	Minimum	Maximum	Standard Deviation
Labor	$18.37	$7.01	$41.40	$ 7.35
Fringe Benefits	3.88	0.67	11.48	2.09
Operating Costs (including vehicle operating costs)	5.10	0.71	14.94	2.99
Overhead Costs	2.78	0.00[a]	10.96	2.16
Depreciation	2.01	0.00[a]	5.70	1.09
Total Cost	$32.08	$13.96	$71.12	$11.53
Fringe Benefits as percentage of Labor Costs	21.9	7.2	56.4	7.3
Vehicle Operating Costs	3.67	.64	12.02	2.32

[a]In one small city, the mayor acted as supervisor of sanitation, so his time was added into direct labor costs, leaving the overhead category void. In several other cases, all vehicles were six or more years old, resulting in a depreciation of zero.

under contract, $669. Although these figures seemingly indicate that private firms pay higher wages than public agencies, the difference is not statistically significant at the 95 percent confidence level. Rather, the difference in average wages paid should be attributed not so much to the public or private nature of the refuse collection organization, but to the uneven geographic distribution of such organizations. The South, with wages generally lower than the national average, has a high proportion of public refuse collection agencies as compared to the Mid-Atlantic region and the Midwest, which tend to have higher prevailing wages and also to have a higher proportion of private firms than average.

A second measure of the level of labor's remuneration relates to the proportion of total labor expenses devoted to supervisory, clerical, and support personnel. It is, of course, possible for a collection firm to operate efficiently, pay prevailing wages, and still incur relatively high costs due to a top-heavy supervisory structure. Direct wages averaged 81 percent of total departmental wage and salary expenses attributable to residential refuse collection; the figure ranged from 60 to 100 percent.

A third measure of labor expenditures is the composition of the fringe benefit package. Again, these data are available only for the 101 communities with municipal collection. Of these, 91 provided social security coverage; 87, workmen's compensation; 95, health insurance; 16, dental insurance; 67, life insurance; 26, disability insurance; 16, unemployment compensation; 88, a retirement fund; 68, uniforms; 60, safety equipment (gloves, for example); and 35, longevity pay.

Table 7-5 presents the cost components for different regions of the country. The data are of interest in determining how applicable the municipal cost component figures are to private refuse collection firms. As Table 7-3 shows, private firms tend to use smaller crews than do the municipal collection agencies; therefore, the relative cost of labor may well be less for the private firm than for the public agency. However, the cities in the municipal sample in the Pacific and Mountain regions use small crews (2.1 and 2.2 men), a size not significantly different from the overall average crew size of private firms.[6] Examination of the cost components of the municipally served cities in the Mountain and the Pacific regions, then, will give an indication of the national cost breakdown of contract-served cities. The table shows that in these two regions, labor costs form a smaller percentage of total costs than they do for all cities: 69 percent of total costs are attributable to wage and fringe benefits on a national basis, but only 56 percent of costs are devoted to these expense categories when crew size is small (2.2 or 2.1 in the Mountain and Pacific regions). These results prevail although fringe benefits are at least as large in the Mountain and Pacific regions as they are in the entire sample. Because the fraction of expenses attributable to labor in the Mountain and Pacific regions was relatively small, the fraction of costs attributable to vehicle and other operating expenditures was greater in these regions than in the nation as a whole. Interestingly enough, vehicle

Table 7-5
Per Household Costs of Residential Refuse Collection, in Cities with Municipal Collection, by Region in Dollars per Year

Cost Component[a]	New England	Mid-Atlantic	East North Central	West North Central	South Atlantic	East South Central	West South Central	Mountain	Pacific
Labor	16.18 (8.58)[a]	17.63 (6.21)	19.84 (7.36)	17.79 (5.45)	24.59 (7.98)	15.77 (6.91)	15.22 (3.60)	13.80 (4.92)	15.70 (5.87)
Fringe Benefits	3.58 (1.56)	3.99 (2.50)	4.51 (1.81)	4.04 (.85)	5.17 (2.61)	2.65 (1.51)	2.76 (1.01)	2.69 (1.09)	3.78 (1.95)
Operating Costs (including vehicle operating costs)	3.35 (2.12)	3.77 (2.20)	3.60 (1.97)	6.59 (2.89)	6.17 (2.75)	3.40 (2.07)	5.41 (2.76)	9.76 (4.36)	6.11 (4.00)
Overhead Costs	1.54 (.93)	2.00 (.91)	2.46 (1.77)	2.18 (.50)	4.46 (2.16)	1.65 (1.69)	2.93 (2.90)	3.27 (3.34)	2.18 (1.15)
Depreciation	1.92 (1.02)	2.25 (1.51)	1.29 (.68)	1.92 (1.50)	2.03 (1.03)	1.48 (.60)	1.81 (.78)	2.97 (1.83)	1.42 (1.45)
Total Cost	26.58 (11.67)	29.65 (10.34)	31.71 (9.92)	32.53 (7.14)	42.43 (12.63)	24.94 (9.27)	28.14 (7.04)	34.48 (5.83)	29.20 (11.32)
Fringe Benefits As Percentage of Labor Costs	22.1	22.6	22.7	22.7	21.0	16.8	18.1	19.5	24.1
Vehicle Operating Costs	2.45	2.85	2.92	3.65	4.88	2.68	3.05	6.41	4.67

See Appendix A for list of states in each region.
[a]Standard deviations in parentheses.

operating costs were the highest in both absolute and percentage terms in these regions as compared to the nation as a whole.

In short, higher wages are paid by the refuse collectors in the Mountain and Pacific regions than by refuse collectors in the remainder of the country, causing a decrease in the use of labor as an input to the collection of solid waste; this reduction of demand for labor is significant enough to reduce the percentage of expenditures attributable to labor. For these regions, then (and, possibly, for all private refuse collection firms), one may conclude that the wage elasticity of demand for collection labor is greater than one. One might also reasonably expect that private firms on average devote a smaller proportion of their refuse collection dollar to labor than do public agencies acting as refuse collectors.

Table 7-6 shows the components of the costs of refuse collection according to city size: cities with populations of less than or equal to 10,000; cities with populations between 10,000 and 50,000; and cities with populations of 50,000 or more. These figures show that as city size increases so too does the magnitude of labor expenses in the refuse collection budget increase. This is the case for fringe benefits as a percentage of wages, and for wages plus fringe benefits as a

Table 7-6
Per Household Costs of Residential Refuse Collection in Cities with Municipal Collection, by City Size in Dollars per Year

Cost Component	City Population		
	⩽10,000	10,000-50,000	50,000+
Labor	18.66	15.13	19.79
	(7.92)[a]	(6.10)	(7.13)
Fringe Benefits	3.81	3.22	4.26
	(2.13)	(1.46)	(2.28)
Operating Costs (including vehicle operating costs)	4.63	4.79	5.72
	(2.61)	(2.41)	(3.61)
Overhead Costs	4.24	1.65	2.30
	(2.54)	(1.06)	(1.72)
Depreciation	2.23	1.81	1.68
	(1.24)	(1.01)	(1.13)
Total Cost	33.58	26.60	33.75
	(13.13)	(9.31)	(10.66)
Vehicle Operating Costs	3.27	3.05	4.28
Fringe Benefits as Percentage of Labor Costs	20.4	21.3	21.5
Wages plus Fringe Benefits as Percentage of Total Cost	66.9	68.9	71.2

[a]Standard deviation in parentheses.

percentage of total refuse collection cost. Overhead costs tend to decrease as city size increases, perhaps because larger cities perform more functions than do small ones, thus having a greater number of departments over which overhead costs may be allocated.

Determinants of the Cost of Refuse Collection

Factors Related to Cost

Previous studies indicate strong relationship between cost of production of residential refuse collection services and the price and quality of labor used in providing the services. As work conditions have an impact on the quantity of labor needed to produce a given level of service, such factors as incentive systems and crew size have been identified as important determinants of the cost of refuse collection. Other determinants of labor's productivity are the characteristics of the capital equipment the labor input is combined with to produce refuse collection services. Work conditions and type of capital equipment available are generally expressed using variables to indicate such factors as vehicle capacity, compaction capability, loading location, age of equipment, crew size, existence of an incentive system, and so forth.

While these determinants of the costs of collection are intuitively comprehensible, there is also an interaction between refuse generation, service levels, and the costs of refuse collection. The greater the amount of refuse generated per household, or the greater the service level provided to each household, the greater is the demand for inputs to the production process and, assuming decreasing marginal productivity of inputs, the greater are the costs of production.

The effect of refuse generation upon input requirements has been explored in a study of the South End of Boston, in which it was found that costs per capita increased significantly with increases in the per capita generation of refuse.[7]

The effect of level of service on labor input requirements has been documented fairly well in several engineering studies of refuse collection. Additionally, cost studies undertaken in relatively homogeneous areas (in which wage and equipment maintenance rates are not expected to vary across the sample) give evidence about the direction of impact of increases in the level of service upon input requirements. Studies based on engineering data invariably show an increase in per capita or household costs (usually within a particular city, so the increase in costs can be attributed to an increase in input requirements) when frequency of collection is increased. A simulation model of the northwest quadrant of Baltimore predicted that increasing the frequency from once to twice a week would result in a 9 percent increase in collection cost, assuming no increase in per capita refuse generation.[8] An actual change in service in a midwestern city from twice-a-week backyard service to once-a-week curbside service was documented by Clark and Gillian as resulting in an almost 50 percent reduction in full-time employees.[9] Finally, a time and motion study of 1,925 collection locations in the South End of Boston revealed significant variations in the minutes of loading time required per stop based on the type of containers used by the occupants.[10] The authors concluded that refuse containerized in disposable plastic sacks yielded a significant enough reduction in loading time (in comparison with loading times for stops where refuse was containerized in cans, supermarket bags, or cardboard boxes) that a municipal program subsidizing the purchase of disposal sacks would probably be economically desirable. Finally, three studies of the cost of refuse collection, each performed on a sample of cities within a state or standard metropolitan statistical area, found significant reductions in per capita cost (and thus, presumably, in labor and equipment inputs) when service was reduced either in frequency or convenience of location (curbside rather than backyard collection).[11]

The set of variables indicating type and use of capital equipment has a less well-documented effect upon the level of labor inputs required to produce a given quantity of a given level of refuse collection services. The effect of crew size on loading time per stop was studied in three California cities by Ralph Stone and Company, Inc.[12] The selected cities had comparable climates, topography, population density, incomes, and containerization regulations. In addition, all three collected weekly from curbside or alley locations. Crew sizes varied from one to three; in the city with one-man crews, the mean man-minutes required to collect one ton of refuse were 26.25 (with standard deviation of 4.64); in the city with two-man crews, mean man-minutes per ton were 43.00 (standard deviation of 9.89); and in the city with three-man crews, mean man-minutes were 63.53 per ton (standard deviation of 7.33). Increases in crew size seem to increase required labor inputs.

Large capacity and the presence of compaction capability reduce the frequency of trips to the disposal site, thus increasing the proportion of paid labor hours actually spent on collection. Conversely more time spent in making a trip to the disposal site reduces collection time per crew shift. Increases in the first two variables reduce the labor inputs required, whereas increases in haul-time would be expected to increase labor requirements. Schreiner, using time and motion data for 24 routes in Oklahoma with three men collecting in rear-loading, 20-cubic-yard compactor trucks providing weekly curbside service, found that for every additional nonroute mile (or haul mile) the number of stops serviced per hour decreased by 1.255 (with a standard error of 0.6779).[13] Conversely, each additional nonroute mile reduced the number of collections per hour by only 0.039 (with a standard error of 0.040) when two-man crews operated front-loading, 24-cubic-yard compactor trucks. Thus, haul-time is statistically significant only for the larger sized crew operating the truck with the smaller capacity.

The age of the collection vehicle is expected to increase required labor inputs for several reasons. First, the older the compactor the less efficient it is likely to be, thus increasing required labor by decreasing the capacity of the vehicle. Second, older vehicles are less likely to have labor-saving features, such as right-hand drives and compactors which operate while the truck is in motion. Finally, older vehicles are more likely to break down in the course of a collection route, thus requiring more labor inputs due to lost time while awaiting repair or arrival of a replacement. A study of loading practices in Chicago found that the pounds loaded per crew hour did decrease significantly with increases in vehicle age.[14]

Finally, work conditions are expected to affect the number of hours of labor input required to provide refuse collection service to a household. While no previous research has documented such a relationship (perhaps due to the fact that where previous studies have obtained data in a consistent manner they concentrated on one city in which neither of these variables would be expected to show any distribution, and that where many cities were studied the lack of consistent definitions of variables led to discarding these variables from the model), it seems reasonable that required labor hours will vary indirectly with the presence of a labor incentive system which requires workers to complete their routes, however long it may take, to qualify for a full day's pay. In the latter system, overtime hours can presumably be reduced, although this effect has not been proved conclusively.

Geographic factors may also affect required labor inputs. The density of collection stops is expected to reduce required labor inputs via a reduction in nonproductive travel time between collection locations. Schreiner found that for 20-cubic-yard, rear-loading compactor vehicles with a three-man crew providing twice-weekly curbside collection services, each additional collection per route mile increased the number of collections per hour by 0.788 (standard error 0.166).[15] Thus, density probably has a positive impact on the productivity of labor.

Increase in annual precipitation would be expected to have two types of effects upon labor's productivity. Because of increases in the weight of refuse presented for collection, tons collected per stop would be expected to increase, but so would loading time per stop. As inputs required are defined in terms of hours of service, precipitation is expected to reduce the productivity of labor while increasing the demand for refuse collection services. That inclement weather does increase collection time has been documented in the study of the South End area in Boston. In this study, it was found that the mean time per stop in rainy weather was 49.3 ±40.3 seconds (196 observations) which differed significantly from the mean time per collection stop in clear weather (36.5 ±34.8 seconds with 1,729 observations.)[16]

These determinants of the quantity and price of labor and of capital goods utilized in the production process determine the costs of residential refuse collection. While the wage rate may be of major importance in determining the total costs of refuse collection, and while the solid waste manager may have no impact on the wage rates prevailing in his area, certain of the other factors that affect labor's productivity and ultimately the costs of collection are under the manager's control and do have a significant impact on the costs of refuse collection. These factors include variables that reflect managerial skill, such as crew size, truck capacity and loading location, routing procedures, absenteeism among workers, use of an incentive system, vehicle management and repair practices. They should be sharply differentiated from factors which are either de jure or de facto beyond the control of the manager. In particular, management has little or no control over prevailing wage rates, quantity of refuse to be collected,[17] city size, density, and weather conditions. In many cities the level of service is determined politically and not by the manager of the refuse collection service. However service levels are determined, the cost of a collection system should be evaluated holding service level constant. Where frequent, convenient refuse collection is provided, service should be considered costly only if it is significantly more expensive than service of the same level provided by other cities.

Efficiency Measures and Cost Determinants

In this section, cities in the sample used to analyze the cost of residential refuse collection are separated by service level provided. Comparisons are made, within a particular service level, of the relative efficiency of cities differentiated by each of the cost determinants within the control of management: crew size, use of an incentive system, absentee rate among collectors and drivers, the capacity of refuse collection vehicles, the loading location of collection vehicles, and the capital intensity of the collection operation. Relative cost data are also presented, again within a given service level, for the same management factors. Relative cost data are of course biased to the extent that prevalence of any one of the managerial options is not equally distributed over the United States, thus, for example, biasing the costs downwards should the option be prevalent in areas with low wages. Efficiency measures do not have the inherent potential for bias that the relative cost measures do, unless other factors, such as climate, have a significant impact on efficiency and, again, unless the use of the management options is not equally distributed across the nation.

Two efficiency measures, both free of reliance upon costs, are used in considering the relationship between management practices and efficiency. The first, direct labor hours per household per year, is equal to the total annual per household number of paid working hours by drivers and collectors. The number will exceed actual working hours in cities where collectors are on an incentive system and generally finish their routes prior to completion of the stated number of hours in their shifts. The number does include overtime hours. It should be stressed that this measure is the sum of all hours for which drivers and collectors are individually remunerated; it is *not* equal to the number of hours for which a *crew* is remunerated (unless, of course, the crew size is one). The second efficiency measure used is households serviced per eight-hour shift. This number is equal to the number of households serviced per week divided by one-eighth the number of paid crew hours per week. (Note that if a community with a thousand households receives twice-a-week service, two thousand households are serviced per week.)

Both efficiency measures are expected to be significantly related to management-determined cost factors. In particular, it is expected that direct labor hours per household per year will increase as crew size increases (as it is expected that additional crew members' productivity, while positive, is decreasing), will decrease with use of a "go-home-when-route-is-completed" incentive system, will increase with increases in the capacity of a collection vehicle (as the crew members will make fewer nonproductive trips to the disposal site and thus spend a greater proportion of their paid working hours actually loading refuse onto the truck), will increase with increases in the use of rear-loading vehicles (side- and front-loading vehicles are more conveniently loaded by the driver, thus encouraging the use of a system in which the driver not only drives, but also loads), and will decrease with increases in the capital intensity of collection (cubic yards of capacity per collection vehicle per crew member). It is expected that households served per crew shift will increase with crew size (although the productivity of an additional crew member is expected to be decreasing, it is always expected to be positive), will increase with the use of a "go-home-when-route-is-completed" type of incentive system, will decrease with increases in absentee rates (this latter expectation would be fulfilled only should the regular labor force not be large enough to provide replacements for absent workers, necessitating the use of below-strength crews or transient, inexperienced laborers), and will increase with increases in truck capacity, the percentage of vehicles loading on the side or at the front, and the capital intensity of the collection operation.

Table 7-7 presents these efficiency measures, disaggregated by service level and by managerially controlled cost determinants. Three service levels are distinguished: once-per-week collection at curb or alley, twice-per-week collection at curb or alley, and once-per-week collection at backyard. Other possible dimensions of service level, such as number of collection receptacles allowed per household, were discarded as inappropriate for this sample since where such limitations existed by statute, they were generally not enforced in practice. Of course, one would expect that as service level increases in quality—either the frequency or the convenience of collection—that the relative efficiency of the

Table 7-7
Mean Values of Efficiency Measures and Cost Determinants

	Efficiency Measures and Service Level[a]					
	Direct Labor Hours per Household per Year			Households per Crew Shift		
Cost Determinants	Once-per-Week Curbside	Twice-per-Week Curbside	Once-per-Week Backyard	Once-per-Week Curbside	Twice-per-Week Curbside	Once-per-Week Backyard
Crew Size						
1 man	2.04 (17)	2.28 (5)	1.63 (4)	274 (17)	318 (5)	135 (4)
2 man	2.73 (48)	3.93 (23)	3.85 (10)	453 (48)	259 (23)	254 (10)
⩾3 man	5.05 (33)	4.99 (38)	6.29 (10)	518 (33)	447 (38)	427 (10)
Incentive System						
yes	3.28 (82)	4.05 (66)	3.57 (23)	471 (82)	372 (66)	359 (23)
no	3.52 (24)	4.45 (11)	7.18 (6)	368 (25)	353 (11)	265 (6)
Absentee Rate						
>7.9%	3.93 (30)	4.38 (33)	5.51 (8)	443 (30)	440 (33)	377 (8)
⩽7.9%	3.22 (49)	4.20 (37)	4.28 (16)	459 (50)	309 (37)	318 (16)
Truck Capacity						
⩽20 cubic yards	4.34 (61)	5.05 (43)	4.92 (18)	421 (62)	402 (43)	308 (18)
>20 cubic yards	1.97 (45)	2.93 (34)	3.53 (10)	482 (45)	328 (34)	415 (10)
Percentage Rear-Loading Trucks						
<90%	2.59 (32)	4.31 (25)	4.06 (12)	408 (33)	354 (25)	410 (12)
⩾90%	3.66 (74)	4.01 (52)	4.50 (17)	464 (74)	377 (52)	290 (17)
Truck Capacity per Crew Member						
<9.1 cubic yards	5.16 (39)	4.98 (41)	2.98 (13)	486 (39)	431 (41)	343 (13)
⩾9.1 cubic yards	2.28 (67)	3.12 (36)	5.68 (15)	424 (68)	299 (36)	349 (15)
Overall	3.33 (106)	4.11 (77)	4.32 (29)	447 (107)	369 (77)	340 (29)

[a]Number of observations in parentheses.

system, measured by direct labor hours per household per year, would decrease. Households served per crew shift are expected to be fewer at backyard than at curbside locations; increases in the frequency of collection, *ceteris paribus*, should increase households served per crew shift as each household will present less refuse per collection, thus decreasing loading time.

As expected, direct labor hours per household per year increase with increases in service level: an average of 3.33 paid crew hours per household per year are incurred if the service is once-per-week curbside whereas once-per-week backyard collection incurs 30 percent more direct labor hours per household per year (DLH). Increases in the crew size result in increases in paid DLHs; were the marginal productivity of the second and the third crew members equal to that of the first, one would expect no increase in DLHs as crew size increased. We conclude, then, confirming the Stone study,[18] that the marginal productivity of additional crewmen decreases for crews in excess of one man. A three-man crew, irrespective of service level, spends over twice the paid labor hours to service a household than does a one-man crew. Some small reduction in DLHs seems to occur when an incentive system is used and when absentee rates are low. Large collection vehicles result in a decrease in DLHs of at least 28 percent (an average of over 40 percent), with the smallest reduction occurring in backyard collection service, for which loading into the collection vehicle represents a smaller proportion of the activities of a crew shift than for curb or alley pickup.[19] Use of trucks that do not load in the rear seems to be associated weakly with a decrease in DLHs. Finally, for curb or alley pickup, the greater the capital intensity of collection, the fewer the DLHs incurred on average. This relationship does not hold for backyard collection, although, as a smaller proportion of the crew shift is spent in loading and dumping activities with this service than with curbside service, one would not expect as significant an impact.

Households serviced per crew shift (HCS) decrease with increases in service levels, as expected. Once-per-week backyard service results in an average of 340 households serviced per crew shift; this is 24 percent less than the number of households serviced per crew shift when the service level is once-per-week curbside. Moving from a one-man to a two-man crew, or from a two-man to a three-man crew, results in an average increase of about 45 percent in the households serviced per crew shift; from a two-man to a three-man crew an approximately equal percentage increase in HCSs is observed. This apparent increasing marginal productivity of labor is probably attributable to the result for the twice-a-week curbside service level category, which, no doubt owing to the small sample size, indicates a decrease in households served per crew shift as crew size is increased from one to two. Given the fact that such an observation does not occur for the other service levels (and, indeed, does not occur when all service levels are aggregated), there seems no reason to reject the indication, from the DLH data, that marginal productivity is positive and decreasing for crews in excess of one.

For all service levels, HCS increases when an incentive system is used. The impact of the absentee rate, as expected, is uncertain. Truck capacity has no clear-cut relationship to HCS, although truck capacity per crew man is clearly negatively related to HCS. This is no doubt due to the fact that as truck capacity per crewman increases, crew size decreases (given the reasonable ranges for truck sizes). Efficiency measured by direct labor hours increased with use of vehicles loading in the front or on the side although efficiency measured by households served per crew shift appears to indicate the reverse relationship—a finding that is no doubt attributable to the fact that rear-loading vehicles are generally staffed by larger crews than are side- or front-loading vehicles, and that, as is seen in Table 7-6, larger crews do result in an increase (though not a proportional increase) in households serviced per crew shift.

Two cost measures may be related to the management practices in a similar fashion. The measures used are collection cost per ton of refuse collected per year (CPT) and the annual cost per household serviced (CPH). Again, the measures are divided according to service level, so the relationship between cost and cost component is considered without distortions due to unequal distributions of service level. These figures are presented in Table 7-8.

Both the cost per ton and the cost per household increase with increases in service level: changing collection location from curb or alley to backyard results in an increase in cost per ton and cost per household of at least 28 percent; increasing the frequency of collection from once to twice per week results in an increase in the cost per ton and cost per household of at least 18 percent.

Crew sizes in excess of one are associated with higher costs, on a per ton or per household basis, than are crew sizes of one. For curb or alley collection, the three-man crew is related, in average, to a lower cost than is the two-man crew but to a higher cost than the one-man crew. This finding is unexpected. Only the cost per household for backyard collection shows the expected increase in cost as the crew size is increased from one man to two men to three men. These figures may be explained by a factor related to cost, such as density. No clear pattern of relationship between use of an incentive system and the cost of collection is seen when averages are based on curb or alley collection locations. The average cost figures for backyard collection do, however, show the expected negative relationship between use of an incentive system and the costs of refuse collection. Absenteeism is not related to the cost of collection in a completely clear-cut fashion, although the cost per ton is less for low absentee rates than for high absentee rates irrespective of service level. The cost per household is not clearly associated with the absentee rate, which is not surprising as a 1 percent increase in absenteeism will increase costs by less than 1 percent, in general, and, given that so few other cost determinants are controlled for in these averages, such a small cost increase may well be outweighed by other factors not controlled for.

The relationship between truck capacity, loading location of vehicles, and capital intensity seems clear. Larger collection vehicles, those with capacity in excess of twenty cubic yards, are associated with costs that range from 2 percent to 41 percent less than the costs incurred when collection takes place using smaller vehicles. Loading location of vehicles seems an important cost determinant. The cost per ton or per household is never significantly greater[20] when the loading location is at the side or front rather than at the rear; in four of the six

Table 7-8
Cost[a] and Cost Determinants

	Cost and Service Level[b]					
	Cost per Ton			Cost per Household per Year		
Cost Determinant	Once-per-Week Curbside	Twice-per-Week Curbside	Once-per-Week Backyard	Once-per-Week Curbside	Twice-per-Week Curbside	Once-per-Week Backyard
Crew Size						
1 man	11.79 (10)	14.69 (2)	28.97 (1)	29.38 (14)	44.06 (5)	26.53 (3)
2 man	26.53 (21)	31.63 (12)	24.48 (5)	31.40 (43)	35.80 (20)	37.61 (10)
⩾3 man	19.46 (24)	25.03 (31)	39.40 (4)	28.33 (30)	33.77 (34)	46.78 (10)
Incentive System						
yes	21.01 (47)	25.69 (41)	22.10 (11)	30.60 (73)	36.12 (58)	35.63 (22)
no	18.12 (13)	27.41 (7)	48.09 (3)	28.48 (23)	31.63 (10)	49.52 (6)
Absentee Rate						
>7.9%	20.32 (24)	27.87 (25)	31.95 (3)	26.08 (27)	33.65 (30)	40.33 (8)
⩽7.9%	19.02 (27)	23.95 (19)	23.06 (9)	30.79 (46)	38.36 (31)	33.19 (15)
Truck Capacity						
<20 cubic yards	21.62 (35)	27.68 (34)	31.51 (8)	33.94 (59)	37.00 (42)	38.21 (18)
⩾20 cubic yards	18.65 (25)	21.71 (14)	22.56 (6)	23.96 (37)	32.97 (26)	37.32 (10)
Percentage Rear-Loading Trucks						
<90%	12.63 (22)	25.95 (17)	17.49 (6)	25.57 (28)	36.15 (24)	33.21 (11)
⩾90%	24.87 (38)	25.94 (31)	35.31 (8)	31.96 (68)	35.08 (44)	42.10 (17)
Truck Capacity per Crew Member						
<9.1 cubic yards	20.31 (26)	27.22 (32)	31.87 (7)	29.74 (36)	35.20 (37)	40.65 (15)
⩾9.1 cubic yards	20.33 (34)	23.39 (16)	23.47 (7)	30.31 (60)	35.76 (31)	36.25 (13)
Overall	20.38 (60)	25.94 (48)	27.67 (14)	30.10 (96)	35.46 (68)	38.61 (28)

[a]All costs adjusted by the ratio of the national average wage to wages in a particular city.
[b]Number of observations in parentheses.

columns of Table 7-8, the cost per ton or cost per household is 25 percent to 100 percent greater in cities where more than nine-tenths of the vehicles are rear loading than in the other cities. As for the number of households served per crew shift, there is no clear relationship between the costs of collection and the capital intensity.

Conclusions

The data presented were drawn from a sample of 315 cities and towns located throughout the United States. It was argued that the technology of refuse collection would depend closely upon the general wage rates prevailing in an area, with more capital intensive technologies likely to be common where wage costs are high. The data support this argument, as well as indicate that refuse collection is characterized by a high degree of substitutability between capital and labor. Where wages are highest, not only is the crew size the smallest and capital intensity the greatest, but also the percentage of refuse collection costs attributable to labor is the lowest. This pattern of substituting capital for labor is followed by public refuse collection agencies. In regions where the greatest substitution of capital for labor by the public agency occurred, the sample was similar to the national sample of private firms. Projecting this result to private firms, one would infer that the significance of labor expenses as a component of total collection costs is less for private firms as a whole than for public agencies as a whole, despite the fact that private firms pay approximately $100 per month more per refuse collector than do public firms.

The finding with respect to the substitutability of capital for labor in refuse collection casts some suspicion on the argument that the costs of refuse collection are proportional to prevailing wage costs. While, clearly, prevailing wage costs have a significant impact on the costs of refuse collection, the importance of other factors in determining the total bill for these services must not be discounted.

One may distinguish two types of factors that affect the total costs of refuse collection: those within the control of the refuse collection manager and those outside his control. Attention in this chapter focused on those factors within the control of the manager. He can influence the technologies with which collection services are produced, but he cannot affect the topography, climate, or physical area over which he must organize a refuse collection effort.

The relative proportions of capital and labor which are desirable in a refuse collection system depend, to some extent, on the service level being provided to community residents. In general, the more convenient the collection location the greater the paid labor input which is required for the collection process. Crew sizes in excess of one are less inefficient where such service is provided than where service is at curbside or alley. Most of the manager-influenced cost determinants showed the expected relationship between efficiency in refuse collection—measured by the number of direct labor hours per household per year and the number of households served per crew shift—and costs of refuse collection—measured by the cost per ton of refuse collected and the cost per household served. In general, efficiency is highest and cost is lowest when the incentive system is used, when labor management procedures ensure a low absentee rate, when the capital intensity (truck capacity per crew member) is high, and where vehicles load in locations (front or rear) convenient to the driver.

Although the factors affecting costs exhibited, in general, the expected relationship between extent of use and resultant costs, not enough of the relevant factors have been held constant in Tables 7-7 and 7-8 to allow formulation of any policy guidelines. In particular, these tables control only for service level as they analyze the relationship between the efficiency and the costs of refuse collection and the managerially determined cost-influencing factors. Clearly, it would be necessary to hold jointly all such relevant factors constant were one attempting to predict the relationship between cost and collection techniques. Such a fine degree of disaggregation would stretch the available sample too far. Thus, the procedures of disaggregation and averaging may not be stretched, with any validity, further than they have been. In the next chapter, the costs of refuse collection are considered in a framework that allows full control of factors related to costs and hence enables the formulation of policy guidelines.

Notes

1. National Commission on Productivity, Solid Waste Management Advisory Group, *Opportunities for Improving Productivity in Solid Waste Collection* (Washington: National Commission on Productivity, 1973), p. 1.

2. Ralph Stone and Company, Inc., *A Study of Solid Waste Collection Systems Comparing One-Man with Multi-Man Crews* (Washington: Government Printing Office, Public Health Service Publication no. 1892, SW-9C, AIN 65, 1969), p. xxvii.

3. Price information was obtained both from the service deliverer and the households serviced by the firm in cities with private and franchise arrangements. Prices reported by the firm were used in all calculations; no systematic difference was found between the prices quoted by the firm and those quoted by the households. This was determined as follows:

Let n equal the number of interviewed households serviced by a particular firm and let k indicate the service level received by the jth household, where $j = 1, \ldots, n$ where n always equals five. Let $HH_j(k)$ be the price for service level k reported by the jth household and let $p(k)$ be the price for service level k reported by the private firm. Then the average discrepancy (DP) between prices reported by households and the firm who services them is:

$$DP = \frac{1}{m} \sum_{j=1}^{m} (HH_j[k]/P[k])$$

where the expected ratio is 1 assuming no bias in reporting of prices. If firms quote low prices, the value of DP will exceed 1: For 49 cities, the average value of DP was 1.01, indicating little grounds for preferring either household reported or contractor quoted fee schedules. The distribution of this ratio is:

Less than 0.85	0.85-0.94	0.95-1.05	1.06-1.15	Over 1.15
0%	25%	59%	6%	10%
0	12	29	3	5

    4. Appendix C contains the survey instruments used to compile these data.

    5. All demographic data are from the 1970 Census of Population.

    6. That is, the difference is not significant at the 95 percent confidence level.

    7. Lawrence J. Partridge and Joseph J. Harrington, "Multivariate Study of Refuse Collection Efficiency," *Journal of the Environmental Engineering Division* 100 (August 1974): 970.

    8. David Gordon Wilson, *The Treatment and Management of Urban Solid Waste* (West Point, Conn.: Technomic, February 1974), p. 73.

    9. Robert M. Clark and James I. Gillian, "Systems Analysis and Solid Waste Planning," *Journal of the Environmental Engineering Division* 100 (February 1974): 22.

    10. Partridge and Harrington, "Multivariate Study," p. 977.

    11. See Robert M. Clark, Betty L. Grupenhoff, George A. Garland, and Albert J. Klee, "Cost of Residential Solid Waste Collection," *Journal of the Sanitary Engineering Division* 97 (October 1971): 567; J.M. McFarland et al., *Comprehensive Studies of Solid Waste Management*, Sanitary Engineering Laboratory, University of California at Berkeley, Report No. 72-3, May 1972, p. 58; Werner Z. Hirsch, "Cost Functions of an Urban Government Service: Refuse Collection," *Review of Economics and Statistics* 47 (February 1965): 92.

    12. Stone and Co., *Study of Solid Waste Collection*.

13. Dean Schreiner, George Muncriff, and Robert Dacis, "Solid Waste Management for Rural Areas: Analysis of Costs and Service Requirements," *American Journal of Agricultural Economics* 55 (November 1973): 571.

14. Jimmie Quon, Masaru Tanaka, and Abraham A. Charnes, "Refuse Quantities and Frequency of Collection," *Journal of the Sanitary Engineering Division* 94 (April 1968): 413.

15. Schreiner et al., "Solid Waste Management," p. 571.

16. Partridge and Harrington, "Multivariate Study," pp. 968-969.

17. Quantity of refuse to be collected is dependent upon factors such as income, size of household, and so forth. In many cities, there is no user charge for the refuse collection services; where there is a charge, it is usually a flat fee rather than a quantity based charge, as discussed in Chapter 6. If households have no choice regarding the fee they pay (whether through property taxes or on a flat-fee basis) then one could not expect refuse generation to be sensitive to the price charged. This condition holds for the municipal and contract cities studied. Refuse generation, or at least volume of refuse collected by a single private firm, is of course dependent upon the price charged by the firm when the household has other collection options. These conditions prevail in the cities with private collection.

18. Stone and Co., *Study of Solid Waste Collection.*

19. For backyard collection, much of the crew's time is spent in hauling the refuse from storage locations to the collection point; this time is unaffected by the use of large collection vehicles.

20. Significance is tested at the 95 percent confidence level.

Service Arrangement and the Cost of Residential Refuse Collection

Barbara J. Stevens

Introduction

This chapter details the reasons why a relationship between service arrangement for refuse collection and the costs of providing such a service might reasonably be expected; it also presents the results of an empirical investigation to test for the existence of such a relationship and suggests some explanations for the findings. The expected interaction of service arrangement and costs of refuse collection are discussed. The next section presents cost data disaggregated by market size and service level and analyzes the relationship between the service arrangement and the cost of collection.[1] The final section explores reasons for the observed differences in costs according to arrangement and draws policy-related conclusions from the findings.

The expectation that the cost of providing refuse collection services will vary with the organizational arrangement depends in part on inherent differences in the behavior of the service deliverer in each different arrangement and in part on the assumption that some economies of scale exist in refuse collection. If economies of scale exist, even if only for very small markets, then clearly there will be some cities for whom city-provided service is more costly than would be regionally provided service, for example. The greater the range of cities over which economies of scale prevail, the more significant this factor in affecting the relative cost of collection provided uniquely to residents of a single town by a public agency as compared to service provided across municipal boundaries by a public or private agency. In short, if economies of scale do exist, at least up to a given level of production, then it is not most cost efficient for a municipality whose size is insufficient to capture these scale economies to produce collection services for its residents. Despite the obvious interrelationship of economies of scale and the identification of the most efficient organizational arrangement, however, the inherent differences of the service deliverer and the impacts of those differences on efficiency will be discussed separately from the issues concerning the existence of scale economies.

Difference Owing to the Identity of the Service Deliverer

Advantages of Government Service Delivery. Perhaps the greatest difference in service delivery occurs when one collector is a public agency and another is a private firm. A municipal or other government provider of refuse collection services may be able to achieve lower-cost production than can a private firm merely owing to its public nature. Three aspects in particular play important roles.

First, municipalities usually are not required to pay income or other taxes; municipalities therefore can purchase inputs to the production process (trucks, fuel, and so on) at a lower cost than can private firms. A private firm, on the other hand, must pay taxes not only on its inputs but also on any profits earned. Information from private firms indicates that taxes of all sorts amount to approximately 15 percent of total revenues. In other words, when customers pay the government $100 for refuse collection services, they receive only those services, but when they pay a private firm $100 they receive not only refuse collection service, but also a bonus of $15 worth of other unidentifiable government services that the firm, in effect, rebates to the customer via its taxes.

Secondly, municipalities can collect taxes and they often provide other utilities, such as water, power, or sewer service. If they use taxes to pay for refuse collection service, or if they bill jointly for two or more services, they are therefore able to collect payment from their customers for refuse collection service at a lower cost than can private firms, except for private firms operating under a contract arrangement.

Finally, municipalities do not earn profits.

For the foregoing three reasons, one might expect that government would be able to deliver refuse collection service at least cost. But despite the obvious advantage that the public agency gains because of its tax-exempt status, its bill-collection ability, and its nonprofit nature, other considerations also inherent to the public/private dichotomy may mitigate or reverse the expectation that the public service deliverer would be the least-cost alternative. These other considerations fall into two basic categories: the absence of the benefits of competition and characteristics of the public enterprise that prevent efficient production.

Monopoly and Competition. When the organizational arrangement for service delivery is one in which the local government grants itself a monopoly, none of the benefits of competition can be obtained. When many firms compete to provide a service, each firm must of necessity strive to produce a better product for a given cost or to provide an identical product at a lower cost than its competitors. In a free market, the organizations that cannot achieve either of these objectives drop out of business, leaving only efficient producers. When there is no competition, but instead only a government monopoly, in the short term there is no incentive for efficient production and no penalty for inefficiency. (Over the long term, of course, if it is inefficient, even a government agency can be replaced by a more efficient alternative, through contracting, reorganization, or revolution, for example.)

If the private sector rather than a government monopoly delivers the service, however, it does not follow that the benefits of competition will be realized. If the industry locally is not competitive and if competition cannot be induced, then service by private firms without government regulation of rates may be the least efficient alternative of all.[2]

On the other hand, there are increases in efficiency to be gained by having one organization service all potential customers. Servicing all households along a route minimizes the negative impact on productivity of low population density by decreasing the nonproductive travel time between pickup points to a minimum. It would be expected, then, that exclusivity will lead to more efficient production of refuse collection services, or, in other terms, to realization of any economies of contiguity which may be present. Exclusivity results when the organizational arrangement is such that either a public agency or a single private firm provides mandatory service to all residents in an area. If the effects of the identity of the service deliverer are small, or, indeed, compensate for each other, then the effects of exclusivity would lead one to expect that either a public or a private firm with an exclusive right to serve a given market would be more efficient than a private firm without an exclusive territory. When the effects of the public/private dichotomy do not counterbalance each other, then, depending upon whether the negative effects of being a public service deliverer are larger (or smaller) than the positive effects, one would expect the municipality as service deliverer to be less efficient (more efficient) than the private firm without an exclusive right to serve a given market.

Note that scale economies may be such that a public agency or a single private firm may provide mandatory service to all residents in only one section of a city and still realize the economies of exclusivity. Therefore, if refuse collection organizations suffer from diseconomies of scale when they are very large, for example, in large cities, they may be subdivided in order to reach a more efficient size without necessarily sacrificing the economies of exclusivity. The latter may continue to be realized by having a single organization, public or private, provide exclusive service in each appropriately sized section of the city.

Inefficiency in Public Enterprises. Government as a service deliverer is unique in that its employees may have permanent civil service status; to the extent that this is the case, rigidities in adjusting most efficiently to changes in conditions in the market may occur. If such rigidities arise, the public service deliverer, despite its advantages of tax exemption and bill collection, may not be the least-cost alternative. When the labor force employed by a municipality has civil service status, desirable changes in the kind of labor used in the production process may be difficult if not impossible to achieve. Even if the civil service status of employees has no effect on the individual productivity of laborers and no effect on the enthusiasm with which managers pursue cost minimization as an objective, constraints on the ability to vary the type and amount of labor input can lead to inefficiencies. When job security also compromises the productivity of the laborer or the supervisor, these inefficiencies would tend to increase. It may be expected, therefore, that collectors who provide refuse collection services utilizing a labor force with permanent civil service status will be less efficient (incur greater costs) than those in which the labor force does not have this status.[3]

To the extent that unionization may produce rigidities similar to those of the civil service, one might expect that refuse collection organizations whose employees are unionized will be less efficient than those whose employees are not. In offering these conjectures about the effect of civil service status and unionization, it must be recognized that efficiency is not the only value to pursue; equity, in the sense of fairness to the employee, is obviously an important value.

In addition to the characteristics of government personnel systems, there are other attributes of government enterprises that may affect efficiency. Niskanen[4] and others have identified budget maximization, rather than cost minimization, as the implicit objective of bureaucracies and bureaucrats. If this is so, it can be expected to lead to inefficiency or overproduction, or both. This latter effect is separable, and will be accounted for by holding constant the level of service.

Diversification. A service deliverer who can diversify into closely related activities may be able to achieve economies owing to joint production. For example, a private firm which collects commercial refuse may be able to use its equipment for residential collection as well; to the extent that joint production of refuse collection services of varying types causes a downward shift in the cost function of the producer, the private firm will gain efficiency. As a producer of diverse services, government can utilize workers in street cleaning, parks maintenance, and general labor categories, for instance, to back up its refuse collection force, increasing its efficiency.

Summary of Hypotheses. If the effects of economies of scale are held constant, the discussion may be summarized in terms of the following hypotheses regarding the effect of organizational arrangement on the efficiency of service delivery:

1. Service delivered by a municipality will be less expensive than service delivered by a private firm, provided that the municipality is at least as efficient as the private firm in producing refuse collection services. The lower costs may be attributed to the fact that the public agency pays no taxes, enjoys the efficiencies of joint billing, and earns no profits.
2. Service delivered by a private firm under contract with a municipality is expected to be less costly than service provided by a private firm contract-

ing directly with the individual, when competition is imperfect. The city has greater bargaining power than does the individual householder and will also ensure that the company serves an exclusive area.
3. Service delivered by a private firm whose rates are regulated by a public agency and who contracts directly with the private householder is expected to be less costly than service provided by a nonregulated private firm contracting directly with the individual householder, when competition is imperfect. Again, this hypothesis depends on the city's successful use of its power in setting rates.
4. Service delivered by a unionized work force will be more costly than service provided by a nonunionized force, as unions will be associated with inefficient work rules.
5. An organization that services all customers will be more efficient than one which services only some of the customers, because the latter arrangement prevents realization of any economies of contiguity.
6. Service provided by a producer with diversified services is expected to be less costly than that provided by the producer of a single service, as the latter has fewer opportunities for realizing economies in areas such as procurements and repairs than does the former.

Economies of Scale

The hypothesis regarding the relative efficiency of alternative organizational arrangements assumes that the size of the market is sufficient to enable realization of any economies of scale which may exist in the production of residential refuse collection service. That some do exist seems quite clear. If, for example, a 20-cubic-yard compactor vehicle can service five hundred households per day, then for a city with once-a-week collection, the vehicle is not being optimally employed unless the city has at least 2,500 households. Additionally, for this city, difficulties in service delivery would arise should its single vehicle break down for a significant period of time. On the other hand, there is no clear-cut evidence that economies of scale do exist to any significant extent once the community reaches a given size (usually large enough to justify the presence of a backup vehicle or vehicles). Most of the economies which may be expected result from a better utilization of capital equipment rather than from improved labor utilization. The latter input can be shifted quite readily to the production of other services such as street cleaning. Aside from increases in the efficiency of capital utilization, increases in the scale of refuse collection organization could yield economies in the procurement of materials (through quantity discounts), in specialization of labor (utilizing preventive maintenance of equipment, performed within the organization, for example) or in separation of functions (scale sufficient to justify the employment of accountants, crew supervisors).

Whether any or all of these potential sources of economies of scale are significant is moot. Little concrete or reliable evidence has been presented in previous studies which proves that there are economies of scale, and if there are, at what level they are fully realized. In fact, the evidence that has been presented leaves in doubt the very existence of any such economies. In their study of refuse collection in California cities, McFarland et al.[5] do attempt to investigate the effect of city size (their entire study deals with cities in which refuse collection services are provided by the municipality) upon the cost of delivering refuse collection services. They state: "Systematic investigation of the structure of costs found in municipal solid waste management indicates that there are moderate economies of scale to be found in the collection plus haul to site or to transfer station function. Economies of scale appear to be exhausted as the population in the service area approaches 30,000."[6] This conclusion is based on the estimation of a total cost function for cities with populations of less than 30,000, less than 50,000, greater than or equal to 30,000, and greater than or equal to 50,000, of the form:

where $C = AQ^v$

C = total cost

A = a constant

Q = tons of refuse collected per year

v = a constant reflecting the existence of economies of scale ($v < 1$ indicates scale economies).

It was observed that v tended to approach a value of 1 as city population approached the category of 50,000 or more. However, these results cannot be accepted at face value because of the authors' failure to control for level of service provided. It is quite possible that smaller cities may have fewer claims on public funds (for example, welfare programs) than larger cities and that consequently smaller cities may be able to provide a higher level of refuse collection services than larger cities. To the extent that this is true (and there is no way of determining this from their published results), economies of scale may not be observable at all. Other studies which purport to determine the city size at which economies of scale are exhausted display similar grievous faults. This is probably attributable more to the lack of accurate data than to the lack of perspicacity on the part of these researchers.

To the extent that significant economies of scale exist, it becomes necessary to tailor hypotheses and policy conclusions for a specified market size.

Classification of Organizational Arrangements

The three major organizational arrangements that will be analyzed may be differentiated according to the degree of public intervention in the production process and the consequent level of economic choices allowed the consumer of the service and its producer. Three different kinds of cities are included in this study: those with municipal collection, where a public agency collects the refuse; those with what we call here *contract collection* (which includes mandatory franchise collection), where a private firm under contract or franchise with the city collects refuse from all residences; and those with what we call here *private collection* (which includes nonmandatory franchise collection), where a private firm collects refuse from some but not all of the residences (the remaining residents haul their own refuse or receive service from a competing firm). (The sample of 315 cities is described briefly in the preceding chapter. For a fuller description of the sample, and a discussion of the data-collection procedure, see Appendixes B and C.)

The next section presents the cost of refuse collection according to organizational arrangement and city size. For the municipal and contract arrangements in the sample, the market served is always equal to the city size. For the private arrangement, the market served is, in general, smaller than the city size, as usually more than one firm is competing for the business of households in the area.[7] Of course, in either the contract or the private arrangement, a firm may service several adjacent communities and treat them as a unit in designing its routes; in this case, the market size may be significantly greater than that of the city itself. The contract arrangement differs from the private in two essential respects: the contract serves an exclusive area whereas the private does not and under the contract arrangement generally only the city is billed whereas under the private arrangement generally each individual household is billed.

Cost and Organizational Arrangements

Table 8-1 presents average cost to the household for all the cities in the sample, stratified by city size and by collection arrangement. The numbers presented are cost to the household per cubic yard and cost to the household per ton. As some cities in the sample had data on quantity of refuse only by volume, others only by weight, and some by both measures, and as conversion factors between yardage and tonnage are notoriously inaccurate, cities were allocated to one or both groups based on data availability. The total number of cities in both groups is 495, because 180 of the 315 cities sampled appear in both groups. It may be seen that mean cost per cubic yard varies from a low of $5.32 (for municipal collection in cities with populations from 10,000 to 50,000) to a high of $8.28 (for private collection in cities with populations in excess of 50,000). Mean cost per ton varies from a low of $18.09 (for contract cities with populations in excess of 50,000) to a high of $30.81 (for private collection in cities with populations in excess of 50,000).

The information in Table 8-1 is useful in that it provides general guidelines concerning the costs of refuse collection. However, in that the figures presented

Table 8-1
Average Cost per Cubic Yard and per Ton for Collection
(Number of cities in parentheses)

	Per Cubic Yard		
Arrangement		Population	
	Under 10,000	10,000 to 50,000	Over 50,000
Municipal	$5.80	$5.32	$7.83
	(32)	(22)	(40)
Contract	6.01	6.49	5.47
	(33)	(38)	(18)
Private	8.15	7.35	8.28
	(46)	(52)	(21)
All	6.84	6.66	7.41
	(111)	(112)	(79)
	Per Ton		
Arrangement		Population	
	Under 10,000	10,000 to 50,000	Over 50,000
Municipal	$22.48	$19.47	$25.87
	(29)	(18)	(46)
Contract	18.86	21.77	18.09
	(10)	(16)	(12)
Private	28.39	23.08	30.81
	(19)	(33)	(10)
All	23.79	21.80	25.22
	(58)	(67)	(68)

are standardized only by city size and service arrangement, they are not suitable for drawing conclusions regarding the impact of even city size and arrangements on costs of refuse collected. As service level, weather and prevailing wages are not controlled for, conclusions could be drawn regarding relative costs only if one were sure that these factors affected all the categories equally. Nevertheless, it should be noted that in every category of city size, whether cost is presented as cost per cubic yard or cost per ton, the mean for private collection is greater than that for either municipal or contract collection. As Table 8-2 indicates, this observation continues to hold when the sample is reduced in size and made more homogeneous by selecting only those cities where service is provided once a week with collection occurring at the curb or alley.

It is not surprising that the average cost per ton or cubic yard is greatest when collection is performed by a firm that does not serve all the households in a city. A firm that services all households in an area is able to reduce

Table 8-2
Average Cost per Cubic Yard and per Ton for Collection in Cities with Once-a-Week Curbside Collection
(Number of cities in parentheses)

Arrangement	Per Cubic Yard		
	Population		
	Under 10,000	10,000 to 50,000	Over 50,000
Municipal	$4.22 (11)	$5.56 (5)	$5.20 (8)
Contract	6.33 (14)	5.13 (20)	6.17 (7)
Private	7.81 (25)	6.88 (25)	7.87 (6)
All	6.60 (50)	6.05 (50)	6.29 (21)

Arrangement	Per Ton		
	Population		
	Under 10,000	10,000 to 50,000	Over 50,000
Municipal	$16.58 (10)	$18.87 (4)	$19.65 (11)
Contract	15.75 (10)	17.35 (16)	13.26 (4)
Private	26.35 (22)	21.02 (31)	30.05 (20)
All	20.94 (22)	19.44 (31)	20.13 (20)

nonproductive travel time between stops to a minimum. When a firm does not service all households along a given route, travel time between stops is necessarily increased beyond the minimum level that could be achieved, given the density of a particular community. The higher costs of private collection could tentatively be attributed to the failure of this arrangement to insure that service to all contiguous residences is provided by one collector or, in other words, the failure of this market structure to capture economies of contiguity. This conclusion must remain tentative, because the figures in Tables 8-1 and 8-2 do not control for density, a factor that could be expected to exert a strong influence on travel time between collection locations.

Other factors besides density and service level have a strong impact on the costs of residential refuse collection. Broadly, these factors may be divided into two categories: those within the control of the manager of the refuse collection operation and those outside his influence. When evaluating the relative costs of different collection systems, it is important to control for the possible impact on costs of those factors that are beyond the influence of the refuse collection manager: prevailing wage rates, quantity of refuse to be collected, city size, density, and weather conditions. In many cities the level of service is considered politically immutable and not truly under management's control, but whether or not service levels can be changed, the efficiency of a collection system should be determined within a given service level category. If a community demands frequent, convenient refuse collection, the community's collection arrangement should be considered costly only if it is significantly more expensive than that of other cities where similar high levels of service are provided.

All of these factors must be held constant when evaluating the relative costs of collection, because each has been shown to impact the cost. Each is discussed in turn.

Wages, comprising the largest component of the costs of refuse collection, are clearly an important determinant of costs. One author has stated that variations in wage rates alone may account for a two to one difference in the per-household costs of collection.[8]

Clearly, there is also a relationship between the amount of household refuse to be collected and collection costs. The greater the total amount of refuse to be collected the greater is the demand for collection resources and, assuming that the productivity of extra labor or vehicles does not exceed that of previously employed units, the greater are the costs of collection. This effect, though intuitively obvious, has been well documented. Partridge found, for example, that an increase in refuse generation resulted in a significant increase in input requirements and thus the cost per household.[9] While increases in total refuse to be collected have a clear, positive impact on costs, it is also likely that the number of collection points necessary to collect a given total quantity of refuse will impact costs. In particular, the fewer the number of stops necessary to collect a given total quantity of refuse, the lower the collection costs would be expected to be, holding other factors constant. Thus, in evaluating the relative cost of collection across municipalities, one must hold constant not only the total quantity of refuse to be collected but also the number of collection locations.

Increased density of collection stops can be expected to reduce the amount of labor required by reducing nonproductive travel time between stops. However, increased density might also mean increased traffic congestion; if this is the case, then, as density—and traffic congestion—increase, the cost of refuse collection may increase instead of decrease.

Finally, weather can affect the efficiency and costs of refuse collection. Precipitation, for example, may have two divergent effects on efficiency. Rain and snow can increase the weight of refuse (if all refuse is not properly containerized) thus causing an increase in tons collected per labor hour, an often-cited measure of collection efficiency. However, increases in the weight of refuse to be collected could also have a negative effect on efficiency by increasing the loading time required per collection stop. Indeed, the latter phenomenon was observed in a study in Boston.[10] Wide variations in temperature would require the collection force to have flexibility for providing service in extremely divergent conditions; where such variation exists it may be necessary to provide two sets of safety equipment, uniforms, and operating procedures (for example, night collection in the summer). The most likely impact of wide temperature variations on costs of collection is, then, positive.

Service level, that is, the frequency and point of collection, is well known to affect costs,[11] as is demonstrated in the preceding chapter. An increase in the frequency of collection or a change in the point of collection from, say, curb or alley to backyard, will increase the cost of collection, if other factors are held constant.

Other factors *under the control of management* also have an impact on the costs of refuse collection. These factors encompass all aspects of labor and equipment management. Indicators of management practices are such factors as absentee rates, the policy regarding incentive systems, crew sizes, and type (rear-loading, side-loading), capacity, and age of collection vehicles. No *one* of these factors identifies a collection system as efficient or inefficient. However, it has been quite clearly demonstrated that most of these measures do affect costs (for example, it has been shown in many cities that smaller crews lead to reductions in the cost of refuse collection).[12] Other factors, such as the presence or absence of a labor incentive system, can be effective or ineffective in reducing costs. If the incentive system utilized is one in which the crew members go home upon completion of work, the system can lead to improved productivity and lower costs or, on the other hand (especially in the absence of good supervision), the system can lead to more missed collections, a higher rate of accidents, and higher costs. In short, although these management indexes are important in determining the efficiency of a collection operation, they are not factors that should be held constant in comparing costs across cities. Rather, by allowing these factors to vary in comparing costs of collection, and by examining differences in management practices between "expensive" and "inexpensive" cities, one can obtain important guidelines regarding factors to be carefully considered in planning an efficient collection system.

Using regression techniques, costs to the household were compared, holding constant those factors beyond the control of management: total amount of refuse collected, amount of refuse per stop, wage rates, population density, weather variation and service level.[13] As was found when only size and service level were held constant (Table 8-2), the costs of private collection were always as great or greater than costs of municipal or contract collection.

Disaggregating the sample by city size revealed that, as expected, some scale economies do exist in refuse collection. Cost to the household decreases for any service arrangement as the market served increases to about 20,000 individuals.

Small cost savings may be achieved for further increases in scale up to about 50,000 individuals. There is no indication that cost per household increases, or, in other words, that diseconomies of scale set in, for markets larger than 50,000 in population, although no city larger than 717,000 was studied.

Table 8-3 shows the relationship between city size and the cost of refuse collection. In general, costs decrease at a decreasing rate as city size increases. The costs of refuse collection in a city whose population is 10,000 are about 85 percent, on a per household basis, of the costs of a city whose population is only 2,500. Full economies of scale are realized when a city's population reaches 50,000. At this point, the costs of collection equal 75 percent of what they would have been, holding all other factors constant, were the city only 2,500 persons in size. Increases in size of market served, then, can account for up to a 25 percent reduction in the costs of refuse collection.

These results indicate that the average cost function for refuse collection is as pictured in Figure 8-1, which shows the cost per household for once-a-week

Table 8-3
City Size and Cost of Refuse Collection

City Size[a] (Population)	Cost of Refuse Collection[b]	Percentage of Initial Cost of Refuse Collection
2,500	30.00[c]	100.0
5,000	26.94[c]	89.8
7,500	25.92[c]	86.4
10,000	25.41[c]	84.7
15,000	24.90[c]	83.0
20,000	24.64[c]	82.1
25,000	24.01[d]	80.0
30,000	23.29[e]	77.6
35,000	23.13[e]	77.1
40,000	22.80[e]	76.0
50,000	22.47[e]	74.9
100,000	22.47[f]	74.9

[a]It is assumed that each person generates one-half ton of refuse per year.
[b]The initial cost of refuse collection is taken as $30.00. The pattern shown on this table is expected irrespective of the service level.
[c]Barbara J. Stevens, *Scale, Market Structure, and the Cost of Refuse Collection*, Columbia University, Graduate School of Business, Center for Government Studies, July 1976, Table 2, sample of ≤20,000 population.
[d]Ibid., sample of <30,000 population.
[e]Ibid., sample of ≤50,000 population.
[f]Ibid., sample of >50,000 population.

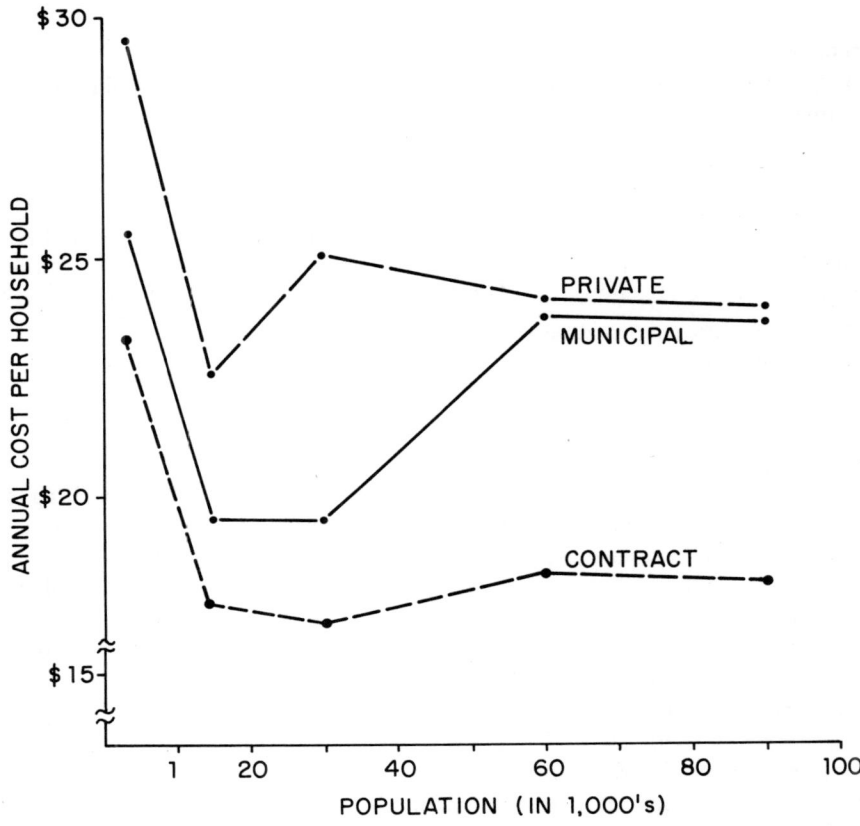

Figure 8-1. Annual Cost per Household for Once-per-Week Curbside Refuse Collection; Estimated Holding Wage Rate, Refuse per Household, Density, Service Level, and Temperature Variation Constant

curbside service. The graph is drawn from the data in Table 8-4, which is calculated from the estimated regression equation using the following reasonable set of values for the factors that must be held constant: Wages equal to $600 per month; density of 600 households per square mile; refuse generation of 1.5 tons per household per year; temperature variation 15 degrees centigrade; size of the average household equal to exactly three persons. These figures are presented for illustrative purposes, to highlight the shape of the cost function and the differences in cost according to arrangement type.

As can be seen from Figure 8-1, cost to the household is always greater under private collection than under any other arrangement, but this difference was found insignificantly different from the costs of municipal collection for small cities (populations up to 20,000) and large cities (populations in excess of 50,000). The cost of private collection was significantly greater than the cost of

Table 8-4
Predicted Annual Cost per Household for Once-a-Week Curbside Collection of Refuse

Arrangement	Population				
	3,000	15,000	30,000	60,000	90,000
Contract	$23.25	$17.77	$17.30	$18.40[a]	$18.26[a]
Municipal	25.51	19.51	19.55	23.77[b]	23.59[b]
Private	29.49[b]	22.55[b]	25.10[b]	24.10[b]	23.92[b]

[a]Significantly different from the cost of municipal collection.
[b]Significantly different from the cost of contract collection.

contract collection in the sample for cities of any population. Firms operating in a competitive environment generally incur billing expenses; firms under contract generally do not have such expenses, because they receive payment directly from the city. Thus, the higher cost of private collection can be attributed to billing expenses and to extra costs due to nonexclusivity within a market area. While no hard data are available to the author to indicate the relative magnitude of billing expenses as a component of the costs of refuse collection, informal discussion with firm owners indicates that approximately 15 percent of total costs may be attributable to billing and collection of fees. If this figure is correct, then the difference in cost between contract and private collection depends approximately equally on differential responsibility for billing expenses and on differences in exclusivity of market areas.[14] This is further discussed in the next chapter.

In cities up to 50,000 in population, the cost of municipal collection did not differ significantly from the cost of contract collection. However, among larger cities in the sample, those with populations in excess of 50,000, those with municipal collection had significantly higher costs than did those with contract collection. In such cities, municipal collection was found to be 29 percent (37 percent) more costly than contract collection when the analysis was restricted only to cities for which data on tons (cubic yards) collected were available. These results, all of which were significant at the 95 percent confidence level or higher, were found not to depend on the presence of either unions or a civil service system.

These results are surprising, because both the municipal agency and the contract firm serve exclusive areas, so neither has an advantage in that respect, while the firm earns profits and pays taxes which the municipal agency generally does not.[15]

This observed difference in cost seems to depend upon differences in management practices and production techniques. As Table 8-5 shows, there are consistent differences in the indexes of management between the contract and municipal arrangements. For cities of all sizes, the contractor uses a significantly

Table 8-5
Management Factors in Refuse Collection

Management Factor	Population of City ≤50,000		Population of City >50,000		Backyard Collection Location	
	Municipal	Contract	Municipal	Contract	Municipal	Contract
Crew size	3.08	2.15	3.26	2.15	3.04	1.98
Truck capacity (cubic yards)	19.04	22.21	20.63	27.14	19.90	23.5
Absentee rate	11%	7%	12%	6%	12%	4%
Percentage of vehicles load- at front and side	26[a]	23[a]	13	44	16	30
Percentage of cities with incentive system	57.1	80.3	80.4[a]	85.7[a]	72.7[a]	86.7[a]

[a]The difference between municipal and contract collection is not significant at the 95% confidence level. All other differences between municipal and contract collection are significant at the 95% confidence level.

smaller crew and a significantly larger collection vehicle than does the municipal collection agency. The absentee rate is significantly higher among public refuse collectors in cities sampled than in private firms. Moreover, these differences increase in magnitude with increases in city size. A further difference in production techniques occurs only in the sample of large cities. In cities with populations greater than 50,000, the private firm under contract with the city is significantly more likely than is the public collection agency to use vehicles that can be loaded conveniently by the driver of the crew, that is, front- or side-loading. These results hold even when cities providing only backyard collection are considered. Even in this case, the private firms use front- and side-loading vehicles almost twice as frequently as do public agencies. In short, whether because of political constraints, or whatever, labor productivity is lower in the municipal refuse collection agencies that were sampled than in the private firms that were sampled, and this difference increases with city size.

Conclusions

The following general policy guidelines can be offered to city officials interested in reducing the cost of refuse collection service to their residents:

1. If it is mandatory that residents receive organized refuse collection service, the cost per household will be least if the city permits only one refuse

collection organization to service an area, provided, if the collector is a private firm, that either rate regulation is of average effectiveness or that contract-letting procedures ensure reasonable competition.
2. Cost to the household decreases under any arrangement as the market served increases to about 20,000 people. Small cost savings may be achieved for further increases in scale up to about 50,000 people.
3. In smaller cities (2,500 to 20,000), municipal collection is no more costly than collection by a private firm—either a firm freely competing for customers or a firm contracting directly with the city.
4. Larger cities (those with populations of 50,000 or more) can expect significantly higher net costs to the household if residential refuse is collected by a public agency or by private firms freely competing for customers than if a private firm, under a contract or a franchise, collects refuse from all residences.

Notes

1. Cost is defined as cost *to the household served*. Where a public agency provides service, costs equal the actual costs incurred by the agency; where a private firm provides the service, cost equals the price paid to the firm by the city or households.

2. If this is the case, then the optimum alternative will be either (a) public-agency performance of the service or (b) setting of rates, or (c) a combination of the two, where a public agency and one or more private firms are service deliverers. The choice will depend upon the relative importance of the positive and negative factors that are likely to affect the relative efficiency of the different alternatives.

3. Should this expectation prove correct, it is not necessarily true that the optimum management alternative is to eliminate civil service protection; the social benefits of employing excessive numbers of employees may be greater than the social costs. In a midwestern city which did decrease its labor force through more efficient management, "it was learned in a follow-up study, which was conducted after the first phase of the labor force reduction, that many of the employees who were laid off did not find comparable employment after a year's time. In fact, many were unemployed a year later." (Robert M. Clark and James I. Gillian, "Systems Analysis and Solid Waste Planning," *Journal of the Environmental Engineering Division* 100 [February 1974]: 22.)

4. William A. Niskanen, *Bureaucracy and Representative Government* (Chicago: Aldine, 1975).

5. Jean M. McFarland et al., *Comprehensive Studies of Solid Waste Management* (Berkeley: Sanitary Engineering Laboratory, University of California, Report no. 72-3), May 1972.

6. Ibid., p. 70.

7. The median number of firms operating in cities with private collection was two.

8. Stephen L. Feldman, "Waste Collection Services: A Survey of Cost and Pricing" in Selma Mushkin, *Public Prices for Public Products* (Washington, D.C. Urban Institute, 1972), p. 228.

9. Lawrence J. Partridge and Joseph J. Harrington, "Multivariate Study of Refuse Collection Efficiency," *Journal of the Environmental Engineering Division* 100 (August 1974): 972.

10. Ibid.

11. See, for example, McFarland et al., *Studies of Solid Waste Management*, p. 71; Clark and Gillian, *Systems Analysis*, pp. 9-10; Werner Z. Hirsch, "Cost Functions of an Urban Government Service: Refuse Collection," *Review of Economics and Statistics* 47 (February 1965): 92; and, Robert M. Clark, Betty L. Grupenhoff, George A. Garland, and Abraham J. Klee, "Cost of Residential Solid Waste Collection," *Journal of Sanitary Engineering Division* 97 (October 1971): 566.

12. Ralph Stone and Company, Inc., *A Study of Solid Waste Collection Systems Comparing One-Man with Multi-Man Crews* (Washington: Government Printing Office, Public Health Service Publication no. 1892, SW-9C, AIN 65, 1969), p. xix.

13. For a technical description of the statistical procedures and results see Barbara J. Stevens, *Scale, Market Structure, and the Cost of Refuse Collection*, Columbia University, Graduate School of Business, Center for Government Studies, July 1976.

14. No significant difference was found between firms providing private collection and firms providing contract collection for any of the management factors to be discussed in greater detail.

15. Indeed, one can attempt to examine the relative efficiency of the two arrangements on an equivalent basis by comparing not the *cost* of municipal service with the *price* charged by the private firm, as was done, but rather by comparing the *cost* of each; the latter can be calculated by subtracting the taxes and profits of the firm.

Industry-wide data on profits are not available, but the published annual reports of several publicly owned firms show that net profits after taxes averaged 5.9 percent of sales; assume a figure of 5 percent. As for taxes, estimates by several people in the industry are that the sum of all fees and taxes paid by private refuse collection firms to all levels of government amount to 15 percent of revenues. Assuming that these estimates are valid, combining them leads to the conclusion that the cost of collection by a private firm is 80 percent of the price it charges. On this basis, the difference of 29 percent (37 percent) cited becomes magnified: the cost of producing municipal collection service is

61 percent (71 percent) greater than the cost of producing service by a private firm, before taxes and profits. (It should be emphasized, however, that tax rates and profit levels were not studied in detail, and this computation is made for illustrative purposes only.)

Local Government Regulation of Residential Refuse Collection by Private Firms

Franklin R. Edwards
and *Barbara J. Stevens*

Introduction

The preceding chapter examined the relative efficiency of the different organizational arrangements for residential refuse collection and concluded that under certain arrangements, in cities larger than 50,000 population, private firms provide less costly service than municipal agencies. The focus of this chapter is on collection arrangements involving private firms. Specifically, we analyze local government regulation of the refuse collection services in seventy-seven cities, all of which utilize private firms for residential collection. Our primary objective is to obtain empirical estimates of the extent to which one regulatory scheme is more efficient than the other. In addition, we examine the effectiveness of the most important regulatory devices used by cities to achieve greater efficiency. Our most significant empirical findings are that substantial cost savings can be realized if government intervenes to collectivize refuse collection service, and that, once this is accomplished, the specific regulatory strategies used make little difference.

The chapter is organized as follows: the next section describes the major collection systems now utilized in this country and points out the main differences among them. Subsequently, we examine the various rationales advanced to support government regulation of refuse collection. The final section presents our empirical findings and discusses their policy implications.

Alternative Collection Arrangements

We can speak of five general types of institutional arrangements for the collection of residential solid waste. At one extreme is public provision of collection services, where municipalities directly provide such services to all households. At the other end of the spectrum is what we call the "unregulated" arrangement, in which private refuse collectors have complete freedom to enter and leave the industry and where collectors and households are free to negotiate whatever refuse service arrangements they wish. In between these two are three general regulatory schemes for private collection: contract, franchise, and license. A detailed description of the institutional arrangements for private collection follows.

In contract arrangements, the city selects and directly pays a collection firm to provide refuse collection services to all residents in a given geographic area. Each collector, therefore, is assured of a monopoly in his area. The typical contract between the city and the collector specifies the level of service to be provided (frequency of pickup, the point of pickup, and the types of refuse to be collected), the payment for each of the services, and the duration of the agreement. (See the next chapter for a detailed discussion of contract terms.) Households are generally given no individual choice regarding the level of service they will receive or the price they will pay—whether payment is *via* a user charge or part of general taxes. (In some cases they may be able to opt for a higher level of service and pay a supplementary fee.) The contract also frequently specifies rate adjustment procedures in the event of changes in the firm's cost, variations in the population served, or increases in some specified price index, and often contains an option to renew the contract if there is "acceptable performance." Among contract cities the most important distinction is whether the city assigns collection contracts by competitive bidding or by simply negotiating directly with would-be collectors. In the empirical work to be reported, we attempt to determine whether this difference is important in practice.

Franchise arrangements are similar but not identical to contract arrangements. The city again gives collection firms an exclusive right to provide service in a specified area for a given period of time, usually sets collection prices, and may or may not utilize competitive bidding (in our sample only three franchise cities made use of competitive bidding). There are, however, two important differences. First, franchised collectors negotiate service levels with households and bill them directly, rather than receiving payment from the city as under the contract arrangement. Thus, franchised collectors bear the costs associated with billing and their collection prices must reflect these additional costs. Second, in some franchise cities it is not mandatory for households to take (and pay for) refuse collection, while in others service is mandatory, as in contract cities. In nonmandatory franchise cities households may self-haul if they are dissatisfied with the collection service being provided, whereas in contract cities self-hauling is not a viable option because all households must pay for collection service whether or not they use it. In our empirical analysis, we assess the practical significance of this self-hauling option. In sum, except for billing costs, mandatory franchise cities should be identical to contract cities, while nonmandatory franchise cities differ because of the self-hauling option.

Under all other private collection arrangements there is no legal agreement between the collector and the municipal government, although some regulation may still exist. In general, these arrangements are characterized by a high degree of competition, easy entry, nonexclusive collection territories, and freedom to set prices and service levels through direct negotiation with service recipients. Two kinds of regulation may exist in these cities, however. In some cities collection firms are required to obtain a hauling license; and in some there is "bureau" price regulation. Since it is possible that a licensing requirement may impose an important barrier to entry, in our empirical analysis we distinguish between a "license" city and all other noncontract, nonfranchise cities—which we refer to simply as "unregulated" cities. Finally, we analyze the effectiveness of price regulation (both by "bureau" and by contract) in all of the arrangements.

In sum, the effectiveness of six distinct regulatory schemes for the provision of the residential refuse collection service are analyzed: contract with competitive bidding; contract without competitive bidding; mandatory franchise; nonmandatory franchise; license; and unregulated. In addition, the effectiveness of price regulation is analyzed.

The Case for Government Intervention

Several arguments have been advanced to support government regulation of refuse collection. A common theme of these arguments is that an unhampered free market will not produce the level and quality of collection services that the public demands, and certainly not at the lowest possible cost. In this section these arguments are briefly described and discussed.

The Public Good—Externality Argument

The first argument for government intervention is that sanitation is a public good, or is characterized by significant externalities. Strictly speaking, a public good is characterized by zero marginal social cost and/or by the inability to exclude additional people from enjoying it, whether or not they contribute to its provision. "Law and order" and "national defense" are such goods. Clearly, refuse collection does not fit this mold. It is a simple matter to exclude noncontributors (or nonpurchasers) from obtaining these services, and marginal social cost is obviously positive. However, sanitation services contribute to another good that may be a public good: public health. (Alternatively, the public good may be a cleaner and more aesthetically pleasing environment.) Given a certain level of public health, however arrived at, communities cannot exclude noncontributors from enjoying its benefits. Individuals who are not willing to pay for the lower risk of disease associated with a cleaner city will nevertheless reap the benefits of a cleaner environment. The possibility of such free riders (or "cheap riders"), therefore, may result in the free market supplying a lower level of sanitation service and public health than is socially optimal. Thus, government involvement with refuse collection is desirable because of the relationship between sanitation service and public health.

Alternatively, sanitation service may be viewed as having significant positive externalities. By purchasing refuse collection service, an individual benefits others as well as himself, without receiving compensation from those so benefited. Thus, in a free-market setting individuals may purchase less sanitation service than is socially optimal, a result similar to that implied by the public good framework.[1]

The Natural Monopoly Argument

Another argument for government intervention is that the economics of the refuse collection industry makes competition structurally impossible. Competition may not be structurally feasible in an industry characterized by important economies of firm size, such as production, capital raising, procurement, and promotional economies.[2] If these economies are large relative to the size of the market, only a few efficient firms can exist. Indeed, the most extreme case of this market condition is where economies of scale are continuous so that only one efficient firm can exist. This condition of natural monopoly is a common justification for government intervention, either in the form of regulation (electric utilities), or through the direct provision by the government of the good in question (subway transportation).

Natural monopoly should be distinguished from markets in which competition is structurally feasible but in which conspiratorial behavior has resulted in monopoly (such as through price fixing, exclusive dealing, tie-in sales, exclusive territorial franchises). These "extra-legal" causes of monopoly are traditionally handled by enforcement of the antitrust laws and are not considered bona fide reasons for either industry regulation or takeover. The preferable policy is to eliminate conspiratorial behavior so that the free market can function effectively, rather than to expunge the market system. Indeed, state antitrust laws have already been successfully applied to the collection industry in a suit against fifty-five refuse haulers in Brooklyn charged with monopolizing the collection industry.[3]

Is the collection industry characterized by natural monopoly conditions? Evidence on economies of scale in refuse collection suggests that there are important economies of scale over small collector size ranges, perhaps up to a maximum of 17,000 to 20,000 households.[4] Consequently, in small cities with populations of less than 50,000 a single collector *may* be the most efficient arrangement. In addition, there are likely to be economies of contiguity[5] in refuse collection, as in any door-to-door service business, such as milk delivery and meter reading. Evidence on the extent of these economies, however, is nearly nonexistent.

Thus, if there is a "natural monopoly" case for government intervention, it would apply primarily to small cities. In large cities, a freely competitive market seems to be both feasible and compatible with industry efficiency, since several efficient refuse firms can exist side by side. Further, although government intervention may be advantageous when a city is small, such involvement becomes less and less necessary as the city grows.

How small must a city be to make government intervention beneficial? The answer to this question depends partly upon the extent of the economies of scale that exist, partly on the costs associated with refuse collectors operating in more than one market (or city) simultaneously, and partly on the nature and extent of the economies of contiguity that might be realized through government intervention. Unlike the traditional natural monopoly industries, such as electricity, the technology of refuse collection does not mandate that a large, indivisible capital investment be made in each market. The basic production unit in refuse collection is the truck, so that the minimum efficient size of market is that which allows use of the most efficient type and number of trucks. On average, one efficient truck can service a city of about 2,500 households, or 7,500 people, once a week. If economies of scale continue through cities of about 50,000, a collector serving six complete cities of 7,500 could conceivably achieve the same economies of scale as could a single collector in a city of 50,000. Viewed in this way, almost no city is small enough to warrant government intervention for the purpose of achieving economies of scale. However, if there are substantial transaction costs associated with operating in more than one city, intervention in the case of small cities may be justified. Finally, if economies of contiguity are significant, and the free market does not result in industry organization that fully captures them, government involvement may again be necessary. In all these cases, of course, whether or not government intervention is beneficial will also depend on the costs associated with such intervention.

The Reliability of Service Argument

A third argument frequently cited to support government intervention is that the free market may not result in stable or reliable collection service. Sanitation services are essential to the everyday functioning of cities. Even short interruptions of service may cause serious public health hazards. Since freely competitive markets are characterized by a considerable amount of exiting and entering activity (as old firms become inefficient and new firms seek to displace existing ones), service disruptions are an obvious possibility. The critical issues are: how reliable is private collection service; how severe (or lengthy) are service disruptions likely to be; and, can government intervention improve reliability?

The evidence suggests that service reliability is not a significant problem in private collection systems. Where service interruptions have occurred, they have been of short duration.[6] There are two reasons for service disruptions: insolvency of the private collector and employee strikes. In general, neither has been a serious problem. When insolvency has occurred, municipalities have found it relatively easy to bring in a new collector.[7] Alternatively, in the case of contract and franchise arrangements, cities have renegotiated collection agreements with failing collectors to place them on a solvent footing.

Strikes are a more serious problem. In response to a strike by private employees, municipalities have several options: bring in a new collector or perform the service themselves, either temporarily or permanently. These possibilities, plus the usual incentives that any private employer has to settle a strike, make it unlikely that strikes by private collection employees will be the cause of major service disruptions.

In any case, even if service continuity were a problem, government intervention is unlikely to improve upon performance in this respect. Even the extreme action of replacing private collection with public collection does not guarantee improved continuity of service. Public employees are striking with increasing frequency[8] and the economic incentives to settle public-employee strikes quickly are probably less than in the case of strikes by private employees. In fact, municipalities are turning increasingly to private firms to provide continued service when their own employees go out on strike.[9] Municipalities are also not immune to bankruptcy.[10] In short, private refuse collection has not been characterized by serious service disruptions, and government intervention seems unlikely to assure service reliability.

The Issue of Organized Crime

A final rationale used to support government intervention is that organized crime tends to dominate and cartelize the industry, making it into a monopoly. This is an intriguing rationale and seems to be unique to the solid waste collection industry (and especially to certain areas of the country).

The argument raises a number of interesting economic issues. Is there something fundamental about the economic structure of the collection industry that makes it a likely target for organized crime? Is the presence of organized crime in the industry a historical accident or a predictable economic phenomenon? Is pervasive public control of an industry the most effective way to eradicate organized crime? Indeed, does (local) government regulation hinder or facilitate the efforts of organized crime to infiltrate and cartelize an industry?

Clearly, careful analysis of these questions is beyond the scope of this chapter. The authors find it hard to believe, however, that there is anything intrinsic to the economic fabric of the collection industry that makes inevitable the presence of organized crime. The best remedy in this industry as in any other would appear to be the vigorous enforcement of our criminal laws, rather than the elimination of the private market system.

The Efficiency of Alternative Regulatory Schemes

This section investigates the empirical question of which private institutional structure for refuse collection is the most efficient in practice. In particular, which of the following six schemes provides collection service to households at the lowest price: contract with competitive bidding; contract without competitive bidding; mandatory franchise; nonmandatory franchise; license; or unregulated? In addition, we analyze the effectiveness of alternative types of price regulation in the collection industry.

Our empirical tests focus on the prices that households pay for service under different collection arrangements. The lower the price, the more efficient the regulatory scheme is judged to be. Alternatively, we consider a regulation to be ineffective if it does not affect the price that households pay.

Data Sample

We utilize data on private collectors operating in 77 cities using some form of private collection system. These data were compiled for the year 1975. In 37 of the cities a contract collection system is used, in 17 a franchise system is used, in 8 a hauling license is required, and in 16 not even a hauling license is necessary, although in one of these cities rates are regulated. Thus, in the sample of 77 cities, private collectors are regulated in all but 15.

The sample is derived from an extensive mail and telephone survey. In specific, 128 contract cities were surveyed, 65 franchise cities, and 125 "noncontract/nonfranchise" cities, some of which turned out to be license and some unregulated cities. To qualify for inclusion in the sample, two factors were required: all collection firms must have provided once-a-week, curbside service, and both collectors and cities must have supplied complete data for all the variables we requested. (This means that both the firm providing service and the city government had to respond to questionnaires.) The former requirement was imposed so that we could test our hypotheses in the context of standardized prices (what households pay for a given, homogeneous service), thereby avoiding the thorny empirical problems associated with analyzing a heterogeneous output. This requirement, of course, reduced our sample considerably. Nevertheless, the sample represents our total survey fairly well. For example, 4 percent of the noncontract and nonfranchise cities in the sample regulate rates, whereas 6 percent of all such cities in the complete survey reported rate regulation.

Lastly, the 77 cities in the sample can be classified into nine categories, as shown in Table 9-1. Six mutually distinct categories appear in the rows, and three others, represented by the columns, cut across these categories. The first

Table 9-1
Number of Cities in the Sample, by Institutional and Regulatory Classification

Institutional and Regulatory Classification	Method of Price Regulation			Total Number of Cities
	Fixed in Contract	Administered by Bureau	None	
Unregulated collection: no hauling license, contract, or franchise is required	0	1	15	16
Licensed collection: hauling license required; no contract or franchise collection	0	3	5	8
Contract collection, with at least two bidders reported	12	0	0	12
Contract collection, other than above	23	1	0	24
Franchise collection, with mandatory service	1	3	0	4
Franchise collection, without mandatory service	9	4	0	13
Total	45	12	20	77

two categories, unregulated and licensed collection, are subclasses of the "private" arrangement that was defined and discussed in earlier chapters, while the next two categories divide the contract and franchise arrangements into subclasses.

Empirical Results

With six distinct collection schemes plus the possibility of price regulation, it is clear that there are a large number of alternative hypotheses to be tested with respect to how each of the collection arrangements differ from one another. In all, therefore, there are 17 distinct hypotheses to be tested—15 collection arrangement hypotheses and two price regulation hypotheses.

Least-squares regression was used to analyze the empirical data and to calculate estimates that were used to test the various hypotheses. Technical details appear elsewhere,[11] but the following principal conclusions can be drawn:

1. The collection schemes fall into two distinct groups, with both forms of contract collection (with and without competitive bidding) in one group and

unregulated, licensed, and nonmandatory franchise collection in the other.[12] (Mandatory franchise collection is discussed separately later.) Collection prices in the latter group of cities are significantly higher than in the first group, corroborating the finding presented in Chapter 8 that contract collection is the most efficient arrangement involving private firms.

2. Mandatory franchise collection cannot be distinguished from either of the preceding two distinct groupings at an acceptable level of statistical significance, probably owing to the fact that there are only four such cities in the sample.

3. Competitive bidding does not appear to influence collection prices;[13] there are no significant price differences between cities that award refuse collection contracts with competitive bidding and those that do so without competitive bidding, as defined in Table 9-1.

4. Mandatory service appears to have lower prices than nonmandatory service, although the finding is weak in the sense that the difference is significant only at a 15 percent level.

5. Price regulation in general has no effect on collection prices; there was no significant price difference between cities with and without regulated prices. In addition, it makes no difference whether price regulation is done by a regulatory bureau or by contractual agreement.

6. As a group, cities with contract collection had collection prices 43 percent lower than the group of cities with licensed, unregulated, or nonmandatory franchise collection. Conversely, the latter group had prices 75 percent higher than the former. This difference is explored in the next section.

Discussion

Chapter 8 noted the similarities in management practices among private collection firms regardless of the organizational arrangement under which they provide service. Therefore, one must look elsewhere to explain the cost difference found here between cities with contract collection and cities with unregulated, licensed, and nonmandatory franchise collection.

There are three likely sources of the differences: billing cost; economies of scale, given the difference in size that exist between the contract collection cities and other cities in our sample; and economies of contiguity.

Billing Costs. Most of this difference in price level cannot be explained by the fact that contract collectors do not incur billing costs, as do refuse collectors in other cities. It is difficult to estimate precisely the additional costs that contract collectors would incur if cities did not assume their billing expenses. It was shown in Chapter 6 that when local government provides collection services and bills its residents, billing expenses are an average of 3.1 percent of total refuse collection costs. However, private billing costs are likely to be higher because of government's ability to bill jointly for several services and to enforce payment. Indeed, discussions with collectors in franchise cities suggested that billing costs may be as high as 15 percent of total collection costs per household (or about 18 percent of total collection cost per household in contract cities). Such an estimate is consistent with the results here. Since the essential difference between cities with mandatory franchise collection and contract cities is that collectors in the former cities incur billing costs while collectors in the latter do not, any price difference should be due to billing costs. This analysis indicates that prices in cities with mandatory franchise collection are about 24 percent higher than in contract cities, which is somewhat higher but still consistent with collectors' estimates of about an 18 percent cost differential. Taking the upper estimate of 24 percent and adjusting prices in contract cities upward still leaves a substantial price difference between contract cities and the other cities: prices in unregulated, license, and nonmandatory cities remain 41 percent higher than in contract cities (or than in cities with mandatory franchise collection); or, alternatively, prices are 29 percent lower in contract cities than in the other group of cities.

Scale and Contiguity

Previous studies of the relationship between market size and the costs of refuse collection have shown that some economies of scale do exist in the refuse collection industry. Increasing the population served from 10,000 to 41,000, for example, would be expected to reduce average collection prices by about 22 percent.[14] Researchers in refuse collection have generally measured firm size as the "number of people per firm," which is derived by dividing total city population by the number of firms operating in the city. This can be a poor measure of firm size because it ignores multiple-city operations by collectors. Nevertheless, for our sample this procedure yields an average firm size of 10,500 people in the unregulated, licensed, nonmandatory franchise cities (which will, of course, understate firm size if firms in these cities operate in more than one city). In contrast, the average firm size in contract and mandatory franchise cities is 41,000.

Thus, if prices are adjusted downward by 22 percent in the cities with unregulated, licensed, and nonmandatory franchise collection to reflect the maximum possible differences between these cities and contract cities due to scale economies, we find that the prices in the former group of cities are still 10 percent higher than in contract cities. (See Table 9-2.) This, then, is the *minimum* price difference attributable to the economies of contiguity. (If, of course, economies of scale are less than we estimated, economies of contiguity will be even greater.)

Table 9-2
Differences in Average Collection Prices; Effects of Billing Costs, Scale, and Contiguity

Adjustment to Mean Price	Annual Price per Household		Difference in Prices	
	(1)	(2)	(3)	(4)
	Contract	Private and Nonmandatory Franchise	(2) − (1)	$\frac{(2) - (1)}{(1)} \times 100$
Base price, calculated by holding independent variables constant	$20.42	$35.73	$15.31	75%
Increase prices in contract cities by 24% for billing costs	25.32	35.73	10.41	41%
Further adjust prices for effect of size of firm	25.32	27.87	2.55	10%[a]

[a]Cost difference due to presence of economies of contiguity.

Summary

Of the various refuse collection arrangements now in use that involve private firms, these findings suggest that the best scheme is contract collection, where cities contract with private collection firms, make collection service mandatory, grant exclusive rights to a single firm to service all households in all or a designated section of the city, and set prices and service standards in the contractual agreement.

Our major conclusion, therefore, is that government regulation of the refuse collection service provided by private firms is beneficial. It can reduce the community's cost of attaining a given level of public health or environmental cleanliness by from 10 to 41 percent.

This cost savings service can be attributed to two economic factors: economies of scale and economies of contiguity.

In addition, the use of competitive bidding to let refuse collection contracts is neither superior nor inferior to simply negotiating contracts with the firms. Price regulation, in any form, also does not affect collection prices, and neither does requiring collectors to obtain a hauling license.

Notes

1. For a thorough analysis of this argument, see F. Edwards and B. Stevens, "Testing the Efficiency of Alternative Organizational Schemes," in *Evaluating the Organization of Service Delivery: Solid Waste Collection and Disposal*, E.S. Savas and Barbara J. Stevens, Principal Investigators, Report to the National Science Foundation (New York: Center for Government Studies, Columbia University, 1977); and "Relative Efficiency of Alternative Institutional Arrangements for Collecting Refuse: Collective Action vs. the Free Market," Working Paper no. 151, Graduate School of Business, Columbia University, 1976.

2. For an excellent discussion of these economies, see Frederick M. Scherer, "Economies of Scale and Industrial Concentration," in *Industrial Concentration: The New Learning*, edited by Goldschmid, Mann, and Weston (Boston: Little, Brown, 1974).

3. See "City Finds It Can't Police Private Refuse Cartmen," *New York Times*, May 29, 1974; and New York vs. A.B.C. Rubbish Removal Co. et al., Supreme Court of the State of New York, indictment no. S314/73. In addition, we do not consider the "extra-legal" activities of organized crime in cartelizing the collection industry to be justification for either regulation or takeover. Here again, alternate and preferable legal remedies are available. See also discussion at infra. p. 12.

4. American Public Works Association, *Refuse Collection Practice* (Chicago: Public Administration Service, 1966); Jean M. McFarland et al. *Comprehensive Studies of Solid Waste Management* (Berkeley: Sanitary Engineering Laboratory, University of California: Report no. 72-3, May 1972; E.S. Savas et al. *Refuse Collection: Department of Sanitation vs. Private Carting*, Report to City of New York, November 1970; Barbara J. Stevens, *Scale, Market Structure, and the Cost of Refuse Collection*, Columbia University, Graduate School of Business, Center for Government Studies, July 1976.

5. Economies of contiguity are economies derived from the contiguous alignment of customers along a service route. In particular, if a collector must pass by B's home while servicing A's and C's homes, it is less costly for him to service B than it would be for another collector who does not pass by B's home while servicing his collection route.

6. This conclusion is based on the author's extensive discussions with both government officials and collectors.

7. The largest known case of service interruption due to insolvency occurred in Omaha, Nebraska, in the early 1960s. The city had no trouble replacing the bankrupt collector.

8. U.S. Bureau of Labor Statistics, "Work Stoppages in Government, 1958-1968," Report no. 348, 1970; U.S. Bureau of Labor Statistics "Analysis of Work Stoppages, 1972," Bulletin 1913, 1974.

9. See also, "Despite Its Workers' Walkout, Bayonne Appears Calm," *New York Times*, September 25, 1975; "6,195 Municipal Defaults Tallied in United States, and in Most Cases Creditors Got Paid," ibid., September 2, 1975.

10. It may be feasible, of course, for municipalities to require performance bonds in order to indemnify themselves from the costs of a service disruption (finding a new private collector). Performance bonds by themselves do not, however, prevent such disruptions from occurring, although by raising barriers to entry (by requiring more capital) performance bonds may indirectly do so by reducing competition.

11. Edwards and Stevens, "Testing Efficiency of Alternative Organizations."

12. Collection arrangements are considered to be in the same group because two criteria are met: the difference in the price level between these collection arrangements is not significant at the 5 percent level; and each of these arrangements has the same relationship with each of the collection arrangements in another group. For example, contract collection systems with and without competitor bidding are placed in the same group because their coefficients are not significantly different from one another and because both of them have coefficients that are statistically different from the coefficients of cities in the other group. This "strong" condition minimizes the chance of incorrectly placing two arrangements in the same grouping, thereby failing to distinguish between them when we should.

13. In an effort to determine whether this result was in any way due to systematic variations in contractual provisions, we also examined the effect of the following contractual provisions on collection prices: the presence of an escalator clause in the contract to adjust collection prices for changes in costs; the value (total and per household) of the performance bond if required; the number of years since the contract was entered into; and the presence of an option to renew the contract. None of these variables was significant and their inclusion did not change any of the results.

14. Stevens, *Scale, Market Structure, and Cost.*

10 Contracts for Residential Refuse Collection

Bennett C. Jaffee

Introduction

The primary characteristic of the contract and franchise arrangements is that the municipality does not directly provide the collection service but rather arranges for a private firm or firms to do so. The municipality's role is to determine the type and level of service, turn the actual collection over to one or more private firms, and see to it that the specified service is performed.

The contract document serves the vitally important role of embodying all the terms of the collection agreement. It is clearly in the municipality's interest to have the terms of the contract enforced, a task that ultimately may require court intervention. More often, however, the municipality itself will be involved in monitoring the private firm's performance. The municipality may also be involved to varying degrees in administrative functions, such as billing customers, receiving customer complaints, or ruling on requests for rate increases.

Contract and franchise collection have been defined; however, note that a *contract* refers to the document signed by the municipality and the private firm whether the municipality has a contract or a franchise arrangement. This undifferentiated definition will be adequate for our purposes.

General Requirements of the Contract Document

The success or failure, for both the municipality and the private firm, of the contract and franchise systems of collection rests in part on the soundness of the contractual framework that governs the collection service. Since it is the contract document that specifies the firm's obligations, the compensation it will receive, and the enforcement mechanisms the municipality may use, it follows that an effective contract should be a complete contract, one that takes into account all areas of possible conflict or disagreement between the municipality and the firm. (To what extent municipalities have contracts that omit what could be called major or important terms is discussed later in this chapter.) It does not follow, however, that a lengthy standardized contract is the best in all situations. The needs of municipalities vary greatly and many cities and towns may enjoy excellent solid waste collection despite having contracts that could be considered incomplete by other standards. Therefore, the discussion that follows should not be interpreted as recommending that a single model contract be uniformly applied. It merely attempts to draw attention to areas of concern that should be of interest to municipal officials.

This chapter will make only passing reference to bidding specifications and municipal ordinances. Bidding specifications are used only when a municipality seeks to secure a contractor by putting the contract out to bid. They are not used at all if the municipality negotiates its contract. Analysis was carried out concerning the different procedures employed by municipalities to enter into a contract, but bidding specifications per se were not separately analyzed. In the great majority of cases where bidding specifications were used, they were explicitly incorporated into the contract. If the specifications were not incorporated, the contract itself generally repeated the terms. Likewise, where a municipal solid waste ordinance exists, the contract generally states that the private collecting firm is bound by the ordinance provisions. It should be emphasized that our concern is not in what document a particular requirement or obligation for the private collector is found but whether it exists at all. We are interested in the totality of all control mechanisms a municipality may exercise over the private collector. While from a strictly legal point of view there are obvious differences between a contract and separate bidding specifications and ordinances, they are lumped together for the purpose of analysis.

Service Specifications

The first and most essential component of the contract is a comprehensive definition or description of the desired service. Its inclusion is vital both to allow the private firm to plan for the service and to ensure that this scope and level of service is maintained throughout the life of the contract. Service recipients, service types, and service levels should be identified clearly. Specification of service is the first level of control over the quality of service, and the municipality may elect to treat deviation from the designated service as a breach of contract. It should also be noted that specification of service by the municipality is a characteristic that distinguishes a contract or franchise system from a licensing or unregulated system.

Flexibility of the Contract

A difficulty in drafting contracts is dealing fairly and adequately with unforeseen circumstances. A well-constructed contractual framework will include not only a comprehensive specification of service but also provisions delineating procedures for amending or modifying the contractor's objections, as well as any other part of the contract, dictated by changed circumstances. A contractual framework which inflexibly locks either side into impossible provisions will only result in an unwanted breach of contract. The procedures for modification, however, should require that a significant need for the change be shown, as well as a notification period.

Protection

Owing to the nature of the service provided, disruption of solid waste collection for even a short period of time can pose a serious public health hazard. Although such disruptions are rare, the municipality should take steps to protect itself against even such an unlikely possibility.

The contract document itself is protection against deterioration or failure of service to the extent it details the obligations of both parties. However, a contracting firm may find itself in a position where it prefers to breach the contract rather than to continue performance, either because the contract is currently disadvantageous or because of the firm's insolvency. The contractual framework can build in extra protective safeguards, beyond the mere exchange of promises to perform. As a protection against the firm's insolvency the municipality may demand, as a qualification to bid, proof of the firm's financial standing. It may also require the right to examine the firm's books or to have an independent auditor do so, or both. Performance bonds can reduce the cost of failure of service to the municipality regardless of whether it is a "convenience" or an insolvency breach, although posting a bond has the effect of increasing the total cost of service.

The municipality should also insure that, in the case of a failure of service, an alternative source of service is available. A common approach is to give the municipality the right to take over the service itself, using (and paying for) the contractor's equipment. The primary objective of these provisions is to insure the continuity and stability of the collection service. These are necessary elements of a contractual framework if it is to guarantee the same stability of service that in preactivist municipal union days was expected from municipal service.

Performance Control

As distinguished from protection against breach of contract, provisions may be written into the contract that control the quality or level of performance of the collecting firm. For example, a contract can require the contracting firm to maintain a facility for the receipt of complaints and impose a penalty for the failure to respond to complaints within a specified time period. The same general approach can be taken to guarantee that the contracting firm does not use unsanitary equipment during its collection, or use an inadequate disposal site.

The contract should specify *how* the contract is to be performed by the contractor only in those areas where the public interest requires certain practices to be followed. To aid in enforcement, the municipality may choose to designate a particular official, by title, as responsible for monitoring the firm's performance. His identity and responsibilities should be written into the contract. Finally, a fine or liquidated damage provision can aid in enforcement.

Municipal and Citizen Obligations

The complement of requiring the contractor to follow designated practices is requiring the service recipients to follow certain practices, such as using an approved type and size of refuse container and placing it at the proper location. If a given practice is found to be desirable, whether for sanitary reasons or because it reduces the cost of collection, it should be mandated in the contract or in an ordinance. The range of the municipality's obligations, such as paying the contractor regularly and guaranteeing the exclusivity of the contractor's service, should also be recorded in the contract.

Limitations on Liability

The contract should contain the range of legal provisions that insulate the municipality from liability for the misdeeds and damages caused by the contractor. The contractor should be required to hold the municipality harmless from claim or suit and agree to pay damages and penalties the municipality may be legally required to pay as a result of granting the contract. The contractor should be required to carry general liability insurance and, as required by state law, workmen's compensation insurance and motor vehicle insurance.

Rates

There are many possible arrangements that a municipality may elect to use to compensate the private collection firm. Key elements of any contract are the provisions that state how much the contractor will be paid for his services, by whom, how frequently, and whether an adjustment in rates is allowed because of changed circumstances. It is essential that those terms be recorded clearly in the contract.

Summary

The foregoing has not been a comprehensive summary of what provisions should be in a contractual framework. It is, instead, a list of the elements and qualities a well-constructed contractual framework should possess if it is to serve its purpose within a contract or franchise system of collection. The seven categories, specification of service, flexibility, protection, performance control, municipal and citizen obligation, limitation on liability, and rate provisions provide the starting point for an analysis of the contractual framework of contracts and franchises.

Discussion of Findings

Methodology

Cities identified through a telephone survey (see Appendix A) as having contract or franchise collection were contacted and asked to complete a mail survey and to send copies of their contracts and other pertinent documents. (See Appendix C for details.) Telephone interviews were used to hasten responses and to clarify ambiguous information. Of the 190 cities identified, 145 responded, of which 100 sent contract documents as well as the other requested material (bid specifications, ordinances, and so on.) These 100 cities (64 contract and 36 franchise) constitute the sample analyzed here.

Frequency Distribution of Contract Terms

One of the purposes of this study is to determine what kinds of contract terms are found in solid waste collection contracts. We are interested in what terms are included in the contract sample and with what degree of frequency they occur. We are also interested in the content or substance of those terms. For example, what are the particular duties and obligations the collecting firm must fulfill?

Before discussing the observed frequency distribution of contract terms, it is useful to compare the contracts with a model solid waste collection contract distributed by the Office of Solid Waste Management Programs of the U.S. Environmental Protection Administration.[1]

Table 10-1 illustrates the extent to which the contract sample contains terms identical or similar to those found in the model contract. The surprising feature of the contract sample is the relatively low frequency with which many contract terms appear. For example, a no-litter provision appears in only 34 percent of the contract sample and a provision establishing a mechanism for handling customer complaints in only 31 percent. Indeed, only seven contract terms appear in more than 80 percent of the sample. However, if many contracts appear incomplete by the model contract standard, it may be that the model contract, even excluding the terms that could not be identified, attempts to be encyclopedic and includes terms not strictly necessary for effective solid waste collection and disposal. A later section in this chapter will seek to explain differences between contracts and why one contract may be more complete than

Table 10-1
Frequency Distribution of Model Contract Terms in Contract Sample

Number	Model Contract Term	Yes	No	Don't Know[a]
1	Definitions	100%	0%	0%
2	Exclusive right	75	23	2
3	Service	82	18	0
4	Term	100	0	0
5	Minimum service	99	1	0
6	House	56	41	3
7	Litter	34	62	4
8	Approved containers	87	7	6
9	Bundles	87	8	5
10	Special and hazardous containers	1	5	94
11	Collection equipment	48	49	3
12	Office	46	51	3
13	Hauling	50	48	2
14	Title to waste	7	87	6
15	Disposal	62	25	13
16	Charges and rates	64	34	2
17	Location	84	14	2
18	Change in cost of doing business	39	58	3
19	Unusual changes or costs	20	77	3
20	Discontinued service or delinquent accounts	23	66	11
21	Complaints	31	68	1
22	Contractor's personnel	3	92	5
23	Compliance with laws	48	48	4
24	Performance bond	77	22	1
25	Payment bond	1	99	0
26	Indemnity	64	34	2
27	Workmen's compensation	65	33	2
28	Assignment	76	22	2
29	Books and records	23	72	5
30	Bankruptcy	1	80	19
31	Permits and licenses	12	83	5
32	Arbitration	9	89	2
33	Standard of performance	62	37	1
34	Modification	46	49	5
35	Illegal provisions	17	78	5

[a] "Don't Know" means that it is unclear from a reading of the contract whether the term in question is covered by the contract.

another. It is simply noted at this point that most of the contracts analyzed are not nearly as complete as the model contract.

The terms that were actually found, and the number of cities that included the terms in their contracts, are shown in Tables 10-2, 10-3, and 10-4. Just as the contracts differ from the model contract, they differ from each other. Of 42 terms whose presence was recorded on a yes-no basis, only 8, or 19 percent, appear in more than 80 percent of the contract sample. Table 10-5 shows how many of the contracts include these terms.

Apparently many municipalities do not find it necessary to include in their contract documents terms that on the surface seem to be worthy of inclusion. It is possible, although beyond the scope of this study, to determine that particular terms, such as a provision that trucks be watertight or a prohibition on truck spillage or leakage, are covered by state law, or county or municipal health codes. It is equally possible that where the same collection firm has provided solid waste collection for an extended period of time a tacit understanding exists between the firm and the municipality and the latter may not feel compelled to include in a contract provisions that have been carried out for years.

It is nevertheless true that municipal officials should consider whether they wish to include in their contracts provisions of the type analyzed. In specific circumstances and especially when a new contract is about to be entered into with a collection firm which has never before provided solid waste collection for the municipality, particular attention should be given to specifying in as much detail as possible clauses concerning the service area and level, compensation, the contractors' obligations, and enforcement.

Factors That Affect Contract Terms

The Effect of Collection Arrangement. The data presented thus far have not distinguished between contract and franchise collection. However, differences do exist. To document the nature of the difference, an analysis was made of thirteen terms considered punitive, restrictive, or in some way limiting the collector's freedom of action. Also, the presence of two other terms thought to be desirable in a contract or franchise document was tested. The results are illustrated in Table 10-6.

Although franchise documents can be written to be just as restrictive as contract documents, in reality they tend to be written with fewer restrictive terms. Of the terms tested, 5 showed approximately the same incidence in both categories, 6 restrictive terms were more common in contract cities, and only 2 were more common in franchise cities. In addition, a higher percentage of solid waste collectors in franchise cities provide commercial service than in contract cities and a higher percentage of franchisees have contracts guaranteeing them

Table 10-2
Terms Included in Sample of Contracts

A. Background Information	Yes	No	Don't Know[a]
Garbage collected under the contract	97%	3%	0%
Trash collected under the contract	94	6	0
Bulk material collected	22	5	73
Yard trash collected	87	8	5
Single-family dwelling service covered	100	0	0
2-4 unit multiple dwellings covered	97	2	1
Over 4 unit multiple dwellings covered	94	5	1
Commercial service covered	73	25	2
B. Service Area and Level			
Provisions for changing number of households served or scope of service	70	26	4
Frequency of collection prescribed	99	1	0
Place of pickup prescribed	84	14	2
Provision limiting hours of collection	56	41	3
Provision prohibiting littering by contractor	34	62	4
Provision requiring free services for the city	24	71	5
Provision specifying disposal site contractor must use	72	25	3
C. Rates, Option to Renew			
Provision setting residential collection rates	64	34	2
Municipal billing for collection services	23	36	41
Contractor billing for collection services	29	30	41
If city bills, must contractor pay a fee for this service	11	13	76
Provision specifying how delinquent accounts are handled	23	66	11
Provision keying rate schedule to cost of doing business	38	59	3
Option to renew	26	70	4
D. Contractor's Obligations			
Provision that trucks be watertight	44	55	1
Provision concerning overnight storage of trucks	12	86	2
Provision prohibiting the overloading of equipment	7	88	5
Provision prohibiting truck spillage or leakage	50	48	2
Provision specifying how often trucks must be cleaned	22	77	1
Provision requiring facility for consumer complaints	46	51	3
Prescribed time period for resolving consumer complaints	31	68	1
Penalty for failing to resolve complaints in a given time	9	90	1
Provision giving city right to audit contractor's books	23	72	5
Provision requiring compliance with state and municipal laws	48	48	4

Table 10-2 (cont.)

		Yes	No	Don't Know[a]
	Provision requiring performance bond	77	22	1
	Provision requiring minimum motor vehicle insurance	19	76	5
	Provision requiring contractor to carry workmen's compensation	65	33	2
	Subcontracts, leases, or assignments require prior approval	76	22	2
	Contractor must indemnify city for contract related legal damages	64	34	2
E.	*Enforcement Clauses*			
	City may take specified actions in case of breach of contract	62	6	32
	Provision for liquidated damages for breach of contract	27	71	2
	Specified procedure for conflict resolution relating to contract	27	71	2
	Provision for amendment of contract	46	50	4
	Provision providing that if contract term is void, remaining terms sustained	17	78	5

[a]"Don't Know" means it was unclear from reading of the contract whether the term in question is covered by the contract.

exclusive rights to a service area. These are, of course, desirable terms for a solid waste collection firm. Further, the only additional requirements franchisees must meet are not of a burdensome nature. The costs for providing certain free services for a city can no doubt be recouped in the collection rate and the requirement that the franchisee provide his own disposal site could actually mean a cost savings if the site is utilized by other collection firms or individuals. It seems clear that a franchisee is less often restricted or burdened by contract terms than a contractor.

It should be pointed out that franchise cities in the sample include a higher percentage of California cities than the contract cities do. Of the 36 franchise cities, 17, or 47 percent, are in California, while the corresponding figure for contract cities is 11 percent. Therefore, any circumstances peculiar to California, such as comprehensive county or municipal health codes or laws or a strong lobbying group on behalf of solid waste collectors, will somewhat skew the figures, but not, in any case, enough to negate the conclusions reached.

The Effect of City Size

The extreme variability of contract content and length, as indicated by the frequency distributions, has not yet been explained. A reading of the contracts

Table 10-3
Distribution of Contracts by Contract Provision

A. *Starting date of contract (percentage of total contracts)*

1961-67	1968	1969	1970	1971	1972	1973	1974	1975
5%	2%	3%	6%	7%	8%	22%	24%	22%

B. *Service area prescribed in contract*

All residences in municipality	All residences in specified areas	Other	Don't know
81%	5%	13%	1%

C. *Frequency of adjustment to rate schedule for changes in cost of doing business*

Once a month	Once every 6 months	Once a year	Unscheduled	Other	Not applicable	Don't know
1%	1%	14%	21%	6%	54%	3%

D. *Frequency of adjustment for changes in number of households served or scope of service area, if contractor's payment is by lump sum*

Once a month	Once every 6 months	Once a year	Unscheduled	Other	Not applicable	Don't know
12%	7%	7%	9%	8%	55%	2%

E. *Term of contract or franchise (in years)*

Less than 1	1	2	3	4	5	8	10	12	15	16	20
5%	14%	11%	24%	1%	27%	2%	8%	1%	3%	1%	2%

F. *Amount of required performance bond*

$0-$999	$1,000-$4,999	$5,000-$9,999	$10,000-$24,999	$25,000-$99,999	$100,000-$249,999	$250,000-$499,999
39%	7%	11%	15%	16%	8%	4%

G. *Amount of required general liability insurance*

$0-$1,000	$50,000-$99,999	$100,000-$249,999	$250,000-$499,999	$500,000-$999,999
21%	2%	9%	37%	30%

H. *Amount of required general property insurance*

$0-$1,000	$10,000-$49,999	$50,000-$99,999	$100,000-$249,999	$250,000-$499,999	$500,000-$999,999
24%	23%	24%	24%	1%	3%

I. *Amount of liquidated damages for failure to complete collection*

0	$1 to $99	$100 to $199	$200 to $499	$500 to $999	$1,000 to $4,999	$5,000+
75%	17%	2%	1%	3%	1%	1%

Table 10-3 (cont.)

J. *Amount of liquidated damages—"other" type breach*

0	$1 to $99	$100
86%	13%	1%

K. *Circumstances under which contract may be renegotiated*

Cannot be negotiated	Unforeseen circumstances	Other	Don't know
90%	1%	8%	1%

gave the general impression that as city size increased, so did the length and detail of the contract document. It appeared that larger municipalities had longer contracts, more terms placing restrictions or requirements on private collection firms, and higher requirements for insurance and performance bonds.

To test the validity of these impressions, cities were divided into four groups by size, and their contract terms were analyzed. Nine of the terms tested were of a restrictive nature to the private collection firm. They are: disposal site specified; watertight trucks; spillage or leakage prohibited; facility for consumer complaints; city has right to audit contractor's books; workmen's compensation required; prior approval for subcontracts, leases, or assignments; city indemnified for contract-related legal damages; and provision for liquidated damages.

Table 10-4
Additional Contract Information from City Questionnaires

		Yes	No	Don't Know
A.	*Cities with Contract Collection*			
	Do the contracts cover all the households in city	92%	6%	2%
	Does city bill residents for regular collections	44	51	5
	Is contractor selected by open bids and fixed or negotiated terms	76	14	10
	Contract awarded to lowest qualified bidder	71	13	16
	Public given opportunity to contest contract awards	46	27	27
	Any breach of performance under presence contract	1	60	39
B.	*Cities with Franchise Collection*			
	All households required to have refuse collection	52	43	5
	Is franchisee required to pay franchise fee to city	69	26	5
	Is contractor selected by open bids and fixed or negotiated terms	55	40	5
	Public given opportunity to contest franchise awards	52	24	24

Table 10-5
Prevalence of Terms in Sample Contracts

	\multicolumn{5}{c}{*Percentage of Contracts*}
	0-20	*21-40*	*41-60*	*61-80*	*81-100*
Number of Contract Terms	6	13	6	9	8

The data are presented in Table 10-7, and show a clear tendency for larger cities to have these restrictive terms more often than smaller ones. One might speculate that the larger the city, the more likely it is to secure the extensive involvement of an attorney in preparing the contract, with the inevitable consequence being a weightier document.

A second conclusion is that the required amounts of performance bonds, general liability insurance, and general property insurance generally increased

Table 10-6
The Difference between Contract and Franchise Agreements with Respect to Contractual Terms

Contract Term	*Percentage of Contract Cities*	*Percentage of Franchise Cities*
1. Limitation on hours of collection	59	49
2. Penalty for failure to resolve complaints in a given time	15	0
3. Workmen's compensation required	72	49
4. Indemnification to city for contract related damage	67	59
5. Liquidated damages	34	14
6. Provision for changing number of households served or scope of service	65	76
7. Watertight trucks	45	46
8. Prohibition of truck spillage or leakage	52	46
9. Facility for consumer complaints	45	49
10. Prior approval required for subcontracts, leases, or assignments	74	78
11. Rate schedule keyed to cost of doing business	59	59
12. Contractor/franchisee must provide disposal site	12	22
13. Free services required for city	17	35
14. Exclusive rights to service area	67	86
15. Commercial service covered	65	86

Table 10-7
Effect of City Size on Contract Terms

Contract Term	Population of City			
	2,500-4,999	5,000-9,999	10,000-24,999	25,000 or more
1. Disposal site specified	50%	69%	60%	65%
2. Watertight trucks	17	32	50	60
3. Spillage or leakage prohibited	22	42	59	60
4. Facility for consumer complaints	6	37	59	62
5. City has right to audit contractor's books	6	0	27	40
6. Workmen's compensation required	56	42	63	80
7. Prior approval for subcontracts, leases, or assignments	44	63	86	90
8. City indemnified for contract related legal damages	67	47	82	60
9. Liquidated damages provision	11	21	27	38
Average Number of Cities with These Terms	31%	39%	57%	62%

with city size. (See Table 10-8.) This is hardly surprising in light of the larger dollar value of the contracts involved.

Finally a number of other terms were analyzed by city size. As Table 10-9 shows, there seems to be no difference in the manner in which private firms are selected to provide contract collection. However, larger cities employ open competitive bidding less often and employ negotiation more often to secure a franchisee. This may be partly explained by the fact that 81 percent of the franchise arrangement contracts are renewals.

Another interesting relationship is the duration of the contract or franchise as a function of city size. This is shown in Table 10-10. Again, the trend, although not completely clear-cut, is for longer contract or franchise terms to be associated with larger municipalities. Only 28 percent of municipalities with less than 5,000 population and 43 percent of cities with 5,001 to 10,000 population have contract or franchise terms of four or more years. The corresponding percentage for cities over 25,000 population is 65 percent.

The last two contract terms examined by city size indicate that larger cities tend more often to include in their contracts provisions for adjusting rate schedules; that is, their contracts are more flexible with respect to compensation. Table 10-11 displays these differences. This finding may be surprising because flexibility with respect to rate schedules is a feature that collection firms would obviously desire. Possible explanations are that large municipalities wish to guard against service disruptions caused by inadequate compensation and that in view of the longer contracts associated with such municipalities, they have

Table 10-8
Effect of City Size on Insurance Amounts

A. Amount of Required Performance Bond

	Population of City			
	2,500-4,999	5,000-9,999	10,000-24,999	25,000+
$0-$24,999	89%	90%	77%	53%
$25,000-$99,999	11	5	23	20
$100,000-$499,999	0	5	0	27

B. Amount of Required General Liability Insurance

$0-$99,999	28%	37%	23%	15%
$100,000-$249,999	22	5	5	7
$250,000-$499,999	39	53	36	30
$500,000-$999,999	11	5	36	48

C. Amount of Required General Property Insurance

$0-$99,999	78%	89%	72%	60%
$100,000-$249,999	17	11	23	35
$250,000-$499,999	0	0	0	2
$500,000-$999,999	5	0	5	3

Table 10-9
Effect of City Size on Manner of Selecting Private Firm

A. For Contract Collection

	Population of City			
	2,500-4,999	5,000-9,999	10,000-24,999	25,000+
Open bids, fixed terms	64%	60%	63%	68%
Negotiation, no bids	27	30	6	18
Open bids, negotiated terms	9	10	25	5
Other	0	0	0	4
Don't Know	0	0	6	5

B. For Franchise Collection

Open competitive bidding	83%	38%	40%	22%
Negotiation	0	38	40	61
Bidding and negotiation	17	12	20	17
Other	0	12	0	0

Table 10-10
Term of Contract or Franchise (in Years) by City Size

Years	Population of City			
	2,500-4,999	5,000-9,999	10,000-24,999	25,000+
Less than 1	6%	5%	9%	2%
1	28%	21%	14%	5%
2	22%	11%	9%	8%
3	17%	21%	41%	20%
4-8	22%	32%	27%	35%
10-15	0%	11%	0%	25%
16-20	6%	0%	0%	5%

agreed to provide mechanisms by which rates may be increased rather than commit a collection firm to a long-term contract with inadequate compensation. Whatever the explanation(s), this seems to be the trade-off for the inclusion of more restrictive terms and higher requirements for performance bonds and insurance.

Policy Guidelines

The general requirements of a contract document have been discussed. Although a lengthy contract containing all the general requirements may be an effective contract, it does not automatically follow that a relatively incomplete contract means that the quality of solid waste collection will suffer. It does bear repeating

Table 10-11
Percentage of Cities that Include Provisions for Adjusting Rate Schedule, by City Size

Reason for rate adjustment	Population of City			
	2,500-4,999	5,000-9,999	10,000-24,999	25,000+
Change in disposal site	17%	5%	18%	30%
Change in cost of doing business	35%	21%	9%	65%

that the following categories of contract provisions should at least be considered by municipal officials: specification of service, flexibility, protection, performance control, municipal and citizen obligations, limitations on liability, and collection rate. Only specification of service and provision for compensation are absolutely essential in a solid waste collection contract. The others are desirable in most situations.

Perhaps the major concern of municipalities concerning solid waste collection is to guard against interruption of service. In this regard particular attention should be paid to provisions that protect the municipality against this occurrence and protect the municipality's interest in contract disputes. It is significant that legal research was unable to discover a single reported case involving a municipal solid waste contract dispute. This can mean either that serious disputes do not generally arise (which may certainly be the case with many municipalities), or that if they do, both sides believe that the disadvantages of litigation (court and legal fees, interruption of service for the municipality, loss of revenue for the collection firm) outweigh the possible advantages. Because disputes that do arise are settled between the parties out of court, municipal officials are especially well advised to include in their contracts protective terms of the type described: a performance bond, right to examine the firm's books, proof of the firm's financial standing, and an alternative source of service. Liquidated damages or arbitration provisions may also be useful. The point is that terms protective of the municipality's interest place the municipality in a better bargaining position with the collection firm if a dispute does arise.

Summing up, it may be said that there are no hard and fast rules to be followed in drafting a solid waste collection contract. Sensitivity to local conditions and an awareness of the kinds of provisions that may benefit a municipality are the keys to an effective contract.

Note

1. U.S. Environmental Protection Agency, Solid Waste Management Office, *Technical Guides and Model Contract for Collection of Residential Solid Wastes*, prepared by National Solid Wastes Management Association and the Solid Waste Management Office, publication SW-81ts (Washington, D.C., 1971).

11 The Role of the Federal and State Governments

Frank P. Grad

Introduction

Until the mid-sixties, both collection and disposal of solid waste were considered exclusively the function of local government. The problem of the disposal of wastes, particularly domestic wastes, never loomed large in nonurban settings, and the legal recognition of the existence of a problem clearly coincides with the period of increasing urbanization.[1] Yet even increasing urbanization did not elevate solid waste collection and disposal above the level of local concern until it became apparent, a mere few years ago, that solid waste collection and disposal was part of the general problem of the control of environmental pollution. As in the case of air and water pollution, control limited to the local level was seen as inadequate to cope with problems that refused to remain confined within narrow jurisdictional boundaries.[2,3]

As recently as 1963, the Advisory Commission on Intergovernmental Relations had concluded, in an analysis of urban functions, that the collection and disposal of solid wastes had no significant economic, social, or other spillover effects, and could therefore be safely left to local management.[4] Yet, a mere two years later, in 1965, President Johnson delivered a special message on conservation and restoration of natural beauty, in which he called for "better solutions to the disposal of solid waste" and recommended national legislation to assist the states in developing comprehensive solid waste disposal programs and to provide for research and demonstration projects.[5] The Solid Waste Disposal Act of 1965 was passed soon after.[6]

Significantly, the Solid Waste Disposal Act of 1965 was passed as Title II of the 1965 amendments to the Clean Air Act.[7] Significantly, too, as the name of the law itself indicates, it was addressed primarily to the problem of *disposal*, rather than collection. This emphasis was not surprising in 1965, a time of growing environmental awareness, particularly in view of concern over air pollution from waste incinerators, increasing evidence that some large metropolitan areas were running out of landfill space, and that landfills, too, contributed to other environmental difficulties, such as water pollution and land use planning problems.[8] It must be stressed, however, that it was the federal concern with disposal—articulated in the 1965 Solid Waste Disposal Act as amended by the Resource Recovery Act of 1970[9]—that caused the states to get involved in what had previously been a purely local concern. It was the states' response to the federal legislation that resulted in their taking first a more active role in waste disposal and then a more direct interest in the collection of solid wastes. Thus, in terms of developments that still continue, legal controls on the collection of solid wastes by state governments find their stimulus in federal laws related primarily to waste disposal, laws that caused the state to get into the regulation and management of solid wastes generally. It is fair to say that state involvement in the collection of solid waste did not have a significant start until about 1970, when states accelerated their efforts to pass legislation in response to certain grant-in-aid provisions in the 1970 amendments of the federal Solid Waste Disposal Act.[10]

The Impact of Federal Legislation

The Solid Waste Disposal Act of 1965 as Amended

The Solid Waste Disposal Act of 1965, as amended by the Resource Recovery Act of 1970, has probably had a minimal *direct* impact on the management of solid waste collection, though it stimulated state action in the field. The federal law provides primarily for assistance to municipalities and states, for grants for research and development, and for planning.[11] In this, the legislation parallels early laws on water and air pollution which stressed the primary responsibility of state and local government for pollution control, with the federal government assuming the role of provider of technical assistance and the means for research and development.[12]

In addition to a variety of detailed findings as to the causes of the solid waste problems and its adverse effects on the environment and reasserting the primacy of local and state obligation to regulate, Congress also finds that "the problems of waste disposal... have become a matter national in scope and in concern and necessitate federal action through financial and technical assistance and leadership in the development, demonstration, and application of new and improved methods and processes to reduce the amounts of waste and unsalvageable materials and to provide for proper and economical solid waste disposal practices."[13] Among the purposes of the act, moreover, is the development of "a national research and development program for improved management techniques, more effective organizational arrangements, and new or improved methods of *collection*, separation, recovery, and recycling of solid wastes, and the environmentally safe disposal of nonrecoverable residues."[14] That the research effort should also include improvements in collection is also noted in the legislative history of the act.[15]

As indicated, a variety of grants for research and demonstration projects for energy and materials recovery, as well as training grants are authorized.[16] A provision added in 1970 also authorizes grants for state, interstate, and local planning for solid waste disposal.[17] As defined under the act, disposal also includes collection.[18] The planning assistance is intended to develop solid waste disposal plans as part of a regional environmental protection system.

In addition to the grant-in-aid provision of the law, the administrator is required, after consultation with appropriate federal, state, interstate, regional, and local agencies to promulgate recommended guidelines after following a prescribed procedure for publication and opportunity for public comment. These guidelines may be promulgated for solid waste recovery, separation, and disposal systems, as well as for collection. The guidelines are to be consistent with the requirements of public health and welfare, with air and water quality standards, and they are to be compatible with appropriate land use plans.[19]

The impact of the federal guidelines is considerable, and is likely to grow when the full range of guidelines is promulgated. Although the law required the administrator of EPA to promulgate guidelines as soon as practicable after the enactment of the Resource Recovery Act of 1970,[20] the first set of guidelines for thermal processing and land disposal of solid wastes was not issued until August 1974,[21] and a year later no guidelines on collection had been issued. The guidelines have considerable impact on the states because no federal grants for research on resource recovery systems and improved solid waste disposal systems are to be made unless the objectives of the grant comply with them.[22] They also apply with binding force to solid waste management by federal agencies and to solid waste practices on federal lands. This is of particular importance on the disposal side, because in western and southwestern states major open areas used or available for land disposal are federal lands.

Other significant provisions of the federal law require, as a condition of federal grants, that there be a single responsible agency to undertake the project, and that there be a well-developed state solid waste management plan, as well as adequate planning to integrate the project into the planning for the entire area's solid waste management.[23] In consequence, virtually all of the states have adopted new legislation since 1970 to make it possible for them and their municipalities to apply for federal grants under the act.[24] These state laws, though frequently focused on disposal, also began to involve the states in the regulation or management of collection.[25]

The federal law also contains significant encouragement for regional or intergovernmental efforts in solid waste management which have already resulted in somewhat greater reliance on intergovernmental arrangements.[26] Planning grants to agencies that include only one municipality may amount to two-thirds of the cost of planning, while intergovernmental agencies may receive up to three-fourths of the planning costs.[27] So too, research and demonstration grants are 50 percent of cost if the facility covers a single municipality, and up to 75 percent when the project serves an area that exceeds a single local jurisdiction.[28]

Other Federal Legislation

A number of other federal laws, mostly recently enacted, have had an impact on the manner in which solid waste is collected. Generally, they affect the nature of the collection vehicle, in an effort to make it safer and less environmentally destructive.

The earliest of these laws is the Federal Aid Highway Act of 1956 which established weight limits for trucks on the National System of Interstate and Defense Highways.[29] The 1956 weight limits still stand, and although it has been asserted that larger, more efficient collection vehicles could be built,[30] the law will not allow this alternative. The result of these federal weight limitations is to limit the payload of collection vehicles. Since in most places collection vehicles go directly to the place of disposal after completion of their route, the impact of these vehicle weight limitations is pervasive.

Other federal requirements that impinge on the collection of solid wastes include the use of so-called computerized braking systems for air-braked trucks as part of the motor vehicle safety standards promulgated by the Department of Transportation in its regulations.[31] These appear to have caused rather special problems associated with the job of collecting wastes that requires the vehicle to stop and start as often as every thirty seconds, and that requires such vehicle to face difficult off-road conditions when depositing wastes at a landfill.[32]

The regulation of the sources of noise under the Federal Noise Control Act[33] also imposes additional burdens, and therefore additional costs, on solid waste management. Two kinds of regulation are involved. The collection vehicle, like other vehicles, must meet noise emission requirements imposed on the truck chassis. In addition it must also meet standards for the refuse body compaction mechanism.[34] Aside from these regulations on the manufacture of vehicles, they must also comply with state laws and regulations pertaining to the regulation of noise for vehicles on the road.[35] Since trucks are major contributors to highway noise, making a relatively far greater addition than passenger automobiles, state regulation in this area has been growing and may be expected to increase.[36]

The Occupational Safety and Health Act of 1970 (OSHA)[37] affects solid waste collection and disposal in important ways, just as it has affected other fields with a high risk of occupational injury. OSHA seeks to reduce the number and severity of work-related injuries and illnesses by the promulgation and enforcement of occupational safety and health standards applicable nationwide.[38] The coverage of the law is particularly significant to the solid waste field because of its mixed private and public character. OSHA applies to all private employers,[39] and, indirectly, will also protect state and local government employees. Although the law does not expressly cover federal, state, and local employees, it provides federal funding to the states and otherwise encourages them to develop and adopt, subject to the approval of the secretary of the Department of Labor, plans for the state management of the occupational health and safety of the employees within the state.[40] Moreover, the regulations governing construction safety are expressly made applicable to construction work performed under the Federal Clean Air and Solid Waste Disposal Acts.[41] Such state plans must provide standards at least as protective as federal standards—and, in addition, must extend similar protections to state and local employees to the extent possible under state law.[42] By mid-1975, half of the states had had their plans approved, and some fifteen more were awaiting approval.[43] Thus, in four out of five states, state and local employees will be protected by OSHA-equivalent standards. Federal employees are also covered pursuant to a presidential executive order.[44]

The detailed safety standards promulgated under OSHA, as well as the recording and reporting requirements under the act (including the keeping of a log of occupational injuries) have added to solid waste management costs, as they have to other industrial costs. The impact of the law can be gauged by some of the solid waste industry opposition to it,[45] as well as by some of the hundreds of reported administrative determinations of OSHA violations concerning regulations governing truck operations[46] and loading equipment[47] used in solid waste collection, and regulating the excavation of trenches and of "sloping"[48] affecting landfill operations.

Advantageous economic treatment is afforded to solid waste disposal facilities under the Internal Revenue Code, which grants preferential treatment to bond issues of state and local governments for the construction of solid waste treatment plants, in that income derived from them qualifies for tax exemption, free from the regulation and limitation imposed on other industrial development bonds.[49]

Favorable tax treatment may well advance the development of plants capable of burning municipal solid wastes for the generation of heat or power. This development has been encouraged by the Energy Supply and Environmental Coordination Act of 1974, which requires the administrator of EPA to study and report to Congress on new energy sources, including "the burning of municipal solid wastes in power plants."[50]

The Growing Role of State Governments in Solid Waste Collection

The Stimulus of Federal Legislation and Involvement

The impact of federal solid waste legislation on waste collection has been most significant, though indirect, in that the states went into the field soon after the 1965 and 1970 enactments of the federal law, either because the federal law made it advantageous for them to do so or because federal involvement drew clear attention to the fact that solid waste management problems had exceeded the local governments' capacities.

The state's greater involvement was both reflected in and aided by the development of a comprehensive model law by the Council of State Governments as part of its 1972 Suggested State Legislation program.[51] The council's "State Solid Waste Management and Resource Recovery Incentive Act" has been extensively drawn on in the states' development of their own laws. Its model act contains a broad range of regulatory powers, including making and enforcing regulations governing solid waste collections.[52] It authorizes contracts or ordinances for the intramunicipal management of solid waste collection for municipalities not participating in intergovernmental arrangements.[53] In both instances, public rendition of services or services rendered by private haulers under contract to the municipality or intergovernmental agency are authorized,[54] and provision is made for regulation of charges, performance bonds, and other contract terms.[55] The major part of the law is devoted to the regulation of, and environmental safeguards for, the disposal of solid waste, and will be discussed in that context. The model act was developed by the National Symposium on State Environmental Legislation, combining the sponsorship and efforts of the U.S. Council on Environmental Quality, EPA, the Department of the Interior, the Council of State Governments, and representatives of state governors' offices and state attorneys general.[56]

The Early Phase–Prior to the Solid Waste Disposal Act of 1965

Before 1965 the collection of solid waste seemed to be almost of as little concern to state governments as it was to the federal government. For the most part state governments viewed collection as a local concern, to be handled by counties and municipalities, not on a statewide basis. Prior to the Solid Waste Disposal Act of 1965, only two states had developed statewide administrative programs for solid waste management.[57]

However, there was some state legislation before 1965 relating to solid waste management in general and collection in particular. Such legislation was not motivated by the state's desire to regulate the field. Rather, its existence was due to the fact that local governments acquire their authority from the state, not by inherent right.[58] While in many instances a local government would be satisfied to base the authority over solid waste management on the general welfare powers bestowed on it by the state constitution, statute, or municipal charter,[59] in other cases state legislation would authorize the local governments to deal with solid waste management specifically. For the most part, it is to be emphasized, specific reference to solid waste matters in pre-1965 state legislation reflects the states' view that solid waste was to be managed on the local level and does not indicate an interest to treat the problem on a statewide basis.

The earliest statutes related to solid waste dealt with the protection of public health and at first were addressed to the resident who generated the waste. Thus, for example, Minnesota enabled its towns in 1899 to order residents to "remove swill, garbage, ashes ... and other foul, nauseous, and unhealthy stuff" or else the town would do so at the owner's expense.[60]

With population growth and the enlargement of towns and cities, it became clear that local governments would have to become involved. Thus a very common pre-1965 state law provision empowered local governments to provide collection services if they chose to.[61]

Some of this early state legislation already reflected a differentiation for areas of various populations, acknowledging their different needs and capacities. Occasionally rural areas, which have the least problem with solid waste and are also least able to provide services,[62] were not included in the enabling provisions.[63] Similarly, the more urgent problems of large urban areas received special attention. As early as 1923, for example, Indiana's enabling provisions allowing counties and municipalities the option to collect imposed a mandatory collection requirement on "first-class cities."[64] So, too, Milwaukee County was required to provide collection services while other counties in Wisconsin had the option, but not the obligation, to do so.[65]

While the states frequently enabled, and occasionally required, local governments to provide collection services, the decision as to how these services were to be provided, whether by municipally provided collection or by private haulers under contract or franchise arrangements, was usually left to the particular local unit.[66] A 1927 Michigan statute authorized the granting of collection franchises by local governments, if 60 percent of the electorate approved. The franchise's fees would have to be approved by the state utility commission as reasonable.[67]

Beyond the passage of such enabling provisions, state governments rarely took an active role in the supervision of solid waste collection before 1965. Several states provided for the licensing of private haulers,[68] but this was unusual. In addition, several states either granted local governments the specific power to regulate the collection of solid waste[69] or else covered certain aspects directly in the legislation.[70]

Another type of provision included in some state statutes, both before and after 1965, restricts haulers from transporting waste materials into and through designated jurisdictions.[71] Such statutes stem either from a fear of depreciated property values or represent an effort to reduce commercial traffic.[72]

Since state legislation on collection prior to 1965 only rarely imposed standards of performance or created duties, state enforcement mechanisms were rarely needed. A notable exception was a Minnesota statute which required that private haulers collecting under contract for a town put up a surety bond.[73] Kansas made it a misdemeanor to engage in unauthorized collection in a city that provided municipal collection services.[74]

Similarly, the lack of state interest is reflected in the relatively small number of statutes that specifically covered the financing of solid waste collection before 1965. Presumably, most localities which provided service directly or through contracts with private haulers, and which did not have user charges or special taxes, appropriated funds from general revenues. Some enabling statutes empowered localities to charge fees for the service rendered.[75] In other instances funds for solid waste management were provided from increases in the property tax[76] or by allocations from other tax revenues.[77]

There were a few instances in pre-1965 state legislation that transcended the purely local approach to solid waste management and, in a sense, anticipated the direction in which the law and administration of solid waste would move very rapidly after 1965. Colorado, for example, had empowered the creation of regional disposal districts for handling collection and disposal,[78] so as to achieve the type of intergovernmental cooperation that federal legislation would stress.

State Legislation on Collection Since 1965

State Solid Waste Management Statutes, Rules and Regulations, and Administrative Agencies

The decade since passage of the Solid Waste Disposal Act of 1965 has seen a virtual explosion of state legislation relating to solid waste management. An area that had been generally unregulated or, at most, regulated to prevent problems of public health or serious nuisances, became, in the matter of ten years, the subject of comprehensive legislation and administrative control. The states enacted all this legislation because the federal legislation made participation in cooperative programs and the receipt of federal funds conditional upon a state's submitting a solid waste management plan and designating a single agency to be responsible for all aspects of that field.[79]

As a result, forty-eight states have either enacted "Solid Waste Management" statutes or significantly amended other existing legislation to achieve the same result.[80] And the two states which had not yet adopted new legislation by 1976, Louisiana and Utah, nonetheless have undertaken to draft state solid waste management plans and issue rules and regulations through their state public health agencies. In fact, every state (except Wyoming) through the agency designated by its recent legislation or through the exercise of preexisting authority over garbage and refuse,[81] has issued rules and regulations in consequence of the federal stimulus. Wyoming, the sole holdout, is engaged, in mid-1975, in drafting its rules and regulations.[82]

In addition to the adoption of statutes or administrative regulations or both, which forty-seven states have now completed, each state, in accordance with the federal law, has designated one of its agencies as its solid waste management agency for the purpose of coordinating activities and plans, issuing regulations, and dealing with the federal government. These agencies fall into two broad categories—public health and environmental protection agencies. Jurisdiction over solid waste management for state public health departments grew naturally out of their concern over public health problems connected with garbage and refuse.[83] Solid waste jurisdiction in environmental protection agencies represents the modern or progressive approach, bringing all environmental programs—air pollution, water pollution, solid waste, conservation, and so forth—under one umbrella department. In at least a half dozen cases state agency involvement has been a two-stage process, with the earlier existing agency, the state department of health, being given jurisdiction initially. Within several years these states have then reorganized their administration and brought solid waste within the newly created environmental protection agency.[84] Although it is likely that health agencies have generally been less enforcement oriented, it is probably true that agency names and classification are no absolute indicator of the substance of their work. It may be worth noting, however, that the division into these categories is just about even—in twenty-six states the management of solid waste is within state departments of public health or their apparent equivalent,[85] and in twenty-five (including D.C.) it is within environmental protection agencies.[86]

The purposes of the federal legislation—to eliminate solid waste as a source of pollution in the 1965 act and to utilize it as a source of recycleable materials and energy in the 1970 amendments—meant that the primary legislative focus of the law and of federal activity subsequent to its enactment was on disposal. Each of these acts also referred to collection, but these references were included to indicate that collection is related to these other problems, not that it is a matter of primary concern itself. Since many states developed programs to become eligible for federal funds and structured their activities to conform to federal standards, the state legislation and regulation generally reflects the federal program emphasis on disposal.[87]

Because the federal legislation has as yet paid only secondary attention to collection, the states' activities in regard to it have been neither as extensive or uniform as they have been in regard to disposal.[88] Every state has granted the designated state agency power to issue rules and regulations covering disposal, and (counting Wyoming with regulations in draft and to be issued by the end of 1975) all fifty states have prepared such regulations.[89] The picture on collection is not quite as complete.

Authority for the designated state agency to issue rules and regulations for collection has been granted specifically only in forty-one states.[90] The others expressly grant authority to regulate disposal and generally make no mention of collection. Ohio is unique in that its statute expressly *excludes* "the collection of solid waste by a political subdivision or a person holding a franchise or license from a political subdivision" from matters to be covered by state regulation.[91]

Yet while statutory authority to issue collection rules and regulations exists in more than forty states, more than a dozen of these have not exercised this authority in regard to collection, limiting regulations issued so far to disposal.[92] The net result is that only slightly more than half of the states have imposed substantively new standards on the collection and transportation of solid waste prior to disposal.[93] And several of these require little more than that solid waste be removed from residences "before creating a nuisance" and that it be transported in a manner designed "to prevent public health and safety hazards,"[94] which does not impose standards significantly different from those enforced by public health agencies long before the recent interest in pollution control or resource recovery.

Beyond the issuance of rules and regulations, the most significant impact on statewide administration of solid waste management has been manifest in the development of solid waste management plans. The adoption of such a plan is a prerequisite for receipt of federal funding.[95] While the requirement that the state agency adopt such a plan appears only in some twenty statutes,[96] the appropriate agencies in the other states have acted without specific legislative directive to develop such a plan.[97] In some twenty states these plans were built upon those that counties and municipalities were required by statute to submit to the state agency.[98] Though the emphasis in these plans is clearly on disposal, the statutory directive included collection in most cases, either expressly[99] or by virtue of its inclusion in the definition of solid waste management.[100] The preparation of such plans means that both local and state officials must consider current and projected needs of the state for solid waste management and alternative arrangements for meeting them.

The Substantive Impact of Recent State Activity on Solid Waste Collection

It is impossible to know from the statutes and regulations alone what has been the actual effect of the recent explosion of state regulation on the day-to-day manner in which solid waste is collected and processed. Not only does the nature of its involvement vary from state to state, but the degree of enforcement of newly imposed standards also affects their impact. Nevertheless the study indicates regulatory trends that undoubtedly affect the way collection is handled.

The Obligation to Provide Service. The recent interest in solid waste management has called states' attention to the role of counties and municipalities in waste collection. Prior to 1965 the power to provide for collection of solid waste was as often inferred from the general police power as it was specifically granted, and only in a few instances, usually involving large urban areas, did state law expressly require the provision of collection services.[101] In the legislation and regulations since 1965, enabling legislation specifically authorizing counties or municipalities, or both, to provide for solid waste collection has been passed in more than two-thirds of the states.[102] About a dozen states now mandate collection at either the county or municipal level or both.[103] In several cases the solid waste management statute imposes the duty explicitly.[104] In other states the duty arises from the regulation,[105] either expressly, as in North Dakota: "The responsibility for refuse collection must be accepted by the city and the county";[106] or by implication, as in Minnesota: "Each county shall provide for a solid waste management system plan to serve all persons within the county."[107] At least one state, Pennsylvania, in requiring its municipalities to provide collection services, exempts rural communities, recognizing the lesser needs and capacities of these areas.[108]

Though a state's solid waste statute or regulations may not mandate expressly that collection service be provided, certain provisions sometimes imply such a duty. Regulations in eleven states set standards for the minimum permissible frequency of pickup from residential property. Five of these occur in states that specifically require counties or municipalities to provide collection service,[109] while the others apply to states where the enabling acts give the local governments the option of arranging for collection or not.[110] Arizona, the District of Columbia, and South Carolina have a statewide requirement of residential collection at least twice a week,[111] the others weekly. South Carolina also requires that "the collector must furnish a collection schedule to all customers being served."[112]

Regulation that requires a minimum frequency of pickup can affect a community's collection arrangements beyond the mere imposition of a certain level of service. At least two states, Arkansas and Pennsylvania, differentiate in regulations between minimum frequencies required for residential pickup as opposed to commercial and industrial collection. Both specify that residential collection should occur at least weekly and that "collection at commercial, institutional (industrial), and public places shall be performed daily or as frequently as necessary."[113] The existence of different standards for different categories of sources, imposed by state regulation or by local ordinance, could influence a county or municipality to use a variety of arrangements, such as municipal or franchised service for residential properties and licensed private haulers for commercial and industrial.[114]

The Arrangement of Collection Services under the New State Laws. It is probable that these expanded duties and requirements for collection will produce a corresponding increase in the need for providers of services. Two features of collection (and also disposal) services are virtually standard in the recent solid waste statutes and regulations. First, no state statute or regulation contains any directive that local governments must perform these services themselves. On the contrary, three-quarters of the statutes expressly empower the local authorities to provide these services in whatever manner they see fit: they may carry out their responsibilities directly or through agreement or contract with other governmental units or with any other person.[115] Thus while private hauling companies may bear heavier costs owing to stiffer new standards, the statute itself does not exclude any of them from the market.

The second characteristic, again found in some three-quarters of the states, the rest being silent, is a provision for intergovernmental cooperation between local units.[116] This, of course, is a direct result of the federal legislation, which increases federal financial assistance for programs that include more than one governmental unit.

A few statutes contain special provisions on franchises. Oregon, Kansas, and South Dakota refer to franchises as permissible contract arrangements,[117] the latter two specifically limiting them to a ten-year maximum term.[118] Missouri limits the duration of grants of exclusive franchises by municipalities to four years.[119] North Carolina limits franchises to seven years.[120] Michigan permits a much longer duration for grants to franchises, thirty years, but requires approval by 60 percent of the electorate to ratify the grant.[121] But beyond these provisions, there was no significant statewide regulation of franchises per se in the laws concerning solid waste management. New York imposes a five-year limit on municipal contracts for collection, regardless of the nature of the arrangement.[122]

The creation of special districts and special quasi-public corporations, as well as statutory authority for the state to step in to perform the functions of delinquent local governments pertain primarily to disposal. It is worth noting that such arrangements can affect collection practices. The growing emphasis on regional approaches, whether through joint or cooperative local government arrangements or through special pollution control corporations, such as Arizona authorizes, with power over collection as well as disposal,[123] will undoubtedly affect the old established patterns for awarding contracts locally. Such intergovernmental arrangements are encouraged by the federal policy. While most states merely authorize them, several—Connecticut, Rhode Island, and South Dakota—also provide in their solid waste laws for incentives in the form of increased state financial aid for such arrangements.[124]

The Effect of Newly Imposed Standards in the Conduct of Solid Waste Collection. In the ten-year period since the passage of the federal Solid Waste Disposal Act in 1965, solid waste collection, which had drawn virtually no state interest prior to 1965, is now subject to state regulation in more than half of the states. Perhaps nowhere is this development more dramatic than in California, where the solid waste management statute, in an unusual provision, prohibited the state board from regulating purely local matters, most of which relate to collection,[125] yet the board has nonetheless declared all such matters to be of state interest and has issued very detailed regulations governing them.[126]

Beyond the fact that substantive regulations on collection are being issued in most states, such statewide rules are generally considered to set minimum standards which local governments may exceed in their own ordinances.

Two-thirds of the states expressly permit local governments to adopt stricter standards on solid waste management,[127] and only one, Hawaii, indicates specifically to the contrary that local laws are not to be stricter than state standards on matters the state has regulated.[128] The power to enact collection standards stricter than those statewide is granted to county governments in thirty-one states[129] and municipal governments in twenty-seven.[130] Thus this type of state law can obviously impose heavier cost burdens on private haulers, but the full extent of the additional burdens can be known only after the effect of local ordinances and regulations issued pursuant thereto has been evaluated.

Collection regulations vary greatly as to scope and detail. About a dozen states require collection from residences either once or twice a week[131] and South Carolina requires the collector to give a schedule to the residents along his route.[132] Even more common, in more than half of the states there are regulations governing the way solid waste is stored by the occupant prior to pickup[133] and the manner in which it must be transported by the collector either to the transfer station or the disposal site.[134]

Certain regulations on storage add little to the old public health standard of trying to avoid "conditions harmful to public health or which create safety hazards, odors, unsightliness, and public nuisances."[135] More commonly, though, the regulations specify that the containers used for storage must be strong, rust- and leak-proof, with tight-fitting lids to keep insects and rodents out, easily cleanable, and with adequate handles for easy handling.[136] Some specifically include tough disposable bags as well,[137] while others merely allow for other devices that meet the standards.[138]

Arkansas is the only state whose laws explicitly authorize counties and municipalities to order the separation of wastes according to type prior to collection,[139] though Connecticut's recycling authority statute refers to "pre-segregation" of refuse as a technique to be considered.[140] With the development of more sophisticated disposal techniques, and with the emphasis on resource recovery and recycling stimulated by federal legislation, over thirty state laws now reflect growing state involvement in this area.[141] It is very possible consequently, that separation prior to collection will become more common. This could, of course, affect collection arrangements and the role of the general hauler.

Nothing was found in state legislation or regulations that would influence a choice between backyard or curbside pickup.

Regulations on the transportation of refuse require, at a minimum, that the vehicles used for collection and transport be constructed and maintained to prevent spillage or leakage.[142] But generally transportation regulations create additional obligations which, if enforced, could mean extra time or expense for the hauler. Some states require the collection vehicle to be covered at all times except while loading and unloading.[143] Others require that the vehicles be maintained in "good repair,"[144] "cleaned as often as necessary,"[145] and even left emptied every night.[146] Several states also explicitly require the hauler to pick up material accidentally spilled promptly and leave the area suitably cleaned.[147]

Beyond the substantive requirements which could put additional burdens and costs on collectors, there has also been a corresponding increase in administrative requirements with which collectors must deal. More than twenty states have either instituted or authorized the introduction of licensure or permit requirements for collectors.[148] Generally, it is the state agency that is required[149] or enabled[150] to issue permits to qualifying collectors; in some instances though, the duty or option is delegated to the local government.[151] Michigan and Mississippi also require collectors to display their permit numbers in large, legible numbers on the sides of their collection vehicles.[152]

Perhaps the greatest impact on the manner in which collectors operate comes not from any regulation dealing specifically with collection but from one dealing with disposal. Since 1965 every state has, through statute or regulation, either eliminated or set a timetable for the elimination of unlicensed disposal sites. With some minor exceptions, primarily for homeowners who dispose of wastes on their own property or very small volume occasional operators, it will soon be against the law to dispose of solid waste anywhere but at a licensed disposal site, which, in theory, meets strict state standards. The federal and state legislation passed in the last decade has begun to put the private, unregulated "dumping" operation out of business. This will no doubt have major impact on the collection industry. Undoubtedly many haulers will be required to drive farther to reach approved disposal operations. Marginal haulers who have been using local dumps may well be put out of business. This development, probably more than any other, will have the greatest impact on the arrangement and manner of solid waste collection.

Mention has been made of impact of federal noise and weight limits on collection vehicles.[153] This is a growing area of state regulation and could place extra burdens on collectors. It is worth noting that Minnesota[154] grants a 20 to 25 percent allowance above normal axle weight limits for refuse compactor collection vehicles and Mississippi[155] charges a $10 fee for an overweight permit, but beyond these there does not seem to be very much express recognition at the legislative level of the special needs of solid waste collection.

Summary

The passage of the Federal Solid Waste Disposal Act in 1965, and its amendment by the Resource Recovery Act in 1970 has caused a great deal of state legislation and regulatory activity. Although most of this activity was originally addressed to the regulation of solid waste disposal, much of it has had a direct impact on solid waste collection. The great majority of states have now assumed some regulatory power over the collection of solid waste, and more than half have actually imposed standards on the collection and transportation of wastes. In spite of this increased state regulatory activity—in what was previously the sole domain of local government—the impact of state regulation on the manner in which solid waste is collected day to day has not thus far been very great. Most of the states have simply enabled their local governments to engage in solid waste collection and less than a fourth of the states actually *require* local governments to collect. Moreover, state law has thus far had relatively little impact on actual arrangements for the management of solid waste collection. Though most state laws—following the federal leadership—grant authority to local governments to make intergovernmental arrangements for the collection of wastes, only a few have actually done so. So, too, although other phases of this study indicate certain regional preferences for collection by public agencies or contract and franchise arrangements respectively, no basis appears in state law for these preferences. In general, state laws authorize municipalities to render collection services themselves or by contract or franchise arrangement, expressing neither a policy preference nor compelling the adoption of one arrangement over another.

A significant impact on cost and arrangements of solid waste collection which must be mentioned stems from state regulation of disposal in carrying out federal environmental requirements. The shutting down of thousands of small dumpsites that have now become illegal, and the concentration of disposal in a smaller number of approved disposal facilities, has clearly resulted in lengthening the distance from the point of collection to disposal, thereby increasing collection costs. This, in turn, may have a tendency to put marginal haulers out of business. Other laws and regulations that have been adopted to protect the public health and the environment are likely to have a similar effect in raising costs. Though the cumulative impact of such federal and state requirements on costs is likely to be substantial, neither the full costs nor the full benefits may as yet be measurable.

Notes

1. Council on Environmental Quality, *Environmental Quality—First Annual Report*, 105-106 (1970); Committee on Pollution, National Academy of Sciences—National Research Council, *Waste Management and Control* (Report to Federal Council for Science and Technology, "Spilhaus Report") 8-16.

2. "Spilhaus Report," *supra* n. 1 at 15-16 *et seq.*

3. Frank P. Grad, George W. Rathjens, and Albert J. Rosenthal, *Environmental Control: Priorities, Policies and the Law*, 160-163 (New York: Columbia University Press, 1971). Frank P. Grad, "Legislative Provisions: Evaluations and Recommendations," in *Regionalized Solid Waste Management*, University of Massachusetts Technical Guidance Center for Environmental Quality, 1972.

4. Advisory Commission on Intergovernmental Relations (ACIR), *Performance of Urban Function: Local and Areawide*, 190-191 (1963).

5. Special Message to the Congress on Conservation and Restoration of Natural Beauty, February 8, 1965, in *Public Papers of the Presidents of the United States*, Lyndon B. Johnson, 1963 (1966).

6. Pub. L. No. 89-272, 79 Stat. 992 (1965), 42 U.S.C.A. § 3251 *et seq.*

7. Ibid.

8. Interstate and Foreign Commerce Committee, H.R. Rep. No. 899, to accompany S. 306, 89th Cong., 1st Sess. (1965), reproduced in 1965 *U.S. Code Congress. & Admin. News* 3608, 3614-15.

9. Pub. L. No. 89-272, 79 Stat. 992 (1965), as amended by Pub. L. No. 91-512, 84 Stat. 1227 (1970), 42 U.S.C.A. § 3251 *et seq.*

10. Pub. L. No. 91-512, 84 Stat. 1227 (1970), 42 U.S.C.A. § 3251 *et seq.*

11. 42 U.S.C.A. §§ 3253, 3254a, 3254b.

12. Frank P. Grad, 1 *Treatise on Environmental Law* (New York: Matthew Bender, 1973), § 2.03[1], § 3.03[1], § 4.02[3].

13. 42 U.S.C.A. § 3251(a)(6).

14. 42 U.S.C.A. § 3251(b)(3).

15. House Report, *supra* n. 7, *U.S. Code Cong. & Admin. News* at 3614.

16. 42 U.S.C.A. §§ 3253, 3254b.

17. 42 U.S.C.A. § 3254a.

18. 42 U.S.C.A. § 3252(5).

19. 42 U.S.C.A. § 3254c(a).

20. 42 U.S.C.A. § 3254c(b)(1).

21. 40 C.F.R. Pt. 240 and Pt. 241, 39 Fed. Reg. 29328 (1974).

22. 42 U.S.C.A. §§ 3254b(b)(1), 3254b(c)(1).

23. 42 U.S.C.A. §§ 3254a(b)(1), 3254a(b)(2).

24. See text at note 51 *infra*.

25. See text at note 80 *infra*. See also, *Council of State Governments, The States' Roles in Solid Waste Management—A Task Force Report* 24-35 (1973). (Hereinafter cited as COSGO Task Force Report.)

26. 42 U.S.C.A. § 3254. See Regional Plan Association, *Waste Management* 67 (New York: 1968); *Regionalized Solid Waste Management* 63-67, 133-139.

27. 42 U.S.C.A. § 3254a(a).

28. 42 U.S.C.A. § 3254b(c)(2).

29. 23 U.S.C.A. § 127.

30. W. Gregory, "Report on the WEMI Truck Chassis Subcommittee," *Waste Age* (November 1974): 35, 36.

31. Motor Vehicle Safety Standard 121, effective March 1977.

32. Gregory, note 30 *supra*, at 38.

33. Pub. L. No. 92-574, 86 Stat. 1234 (1972), 42 U.S.C.A. § 4901 *et seq.*

34. See, *e.g.*, Transportation Equipment Noise Emission Controls: Proposed Standards for Medium and Heavy Duty Trucks, 39 Fed. Reg. 38338 (Oct. 15, 1974).

35. See Grad, Rosenthal et al., *The Automobile and the Regulation of Its Impact on the Environment* 441-449 (Legislative Drafting Research Fund 1975).

36. *Id.* at 442-443.

37. Pub. L. No. 91-596 (1970), 84 Stat. 1591, as amended by Pub. L. No. 93-237 (1974), 29 U.S.C.A. §§ 651-678.

38. 29 U.S.C.A. § 651, 653.

39. 29 U.S.C.A. § 653.

40. 29 U.S.C.A. §§ 667, 672.

41. 29 C.F.R. § 1926.12.

42. 29 U.S.C.A. § 667(c)(2).

43. Chart on Status of State Plan Activity, BNA *Occupational Health and Safety Reporter* 81:1003 *et seq.* (June 12, 1975).

44. Executive Order No. 11807, 39 Fed. Reg. 35559 (Oct. 2, 1974).

45. See, *e.g.*, G. Van Beek, "Occupational Safety and Health Act," *Waste Age* Oct. 1974 at 10; Same, "Poor OSHA," *Waste Age* May 1975 at 14.

46. 29 C.F.R. 1926.600 *et seq.* See, *e.g.*, Glade Inc. 1 OSCH 3292 (1974).

47. 29 C.F.R. 1926.600(a)(3)(i). See, *e.g.*, Glade Inc. 1 OSCH 3292 (1974).

48. 29 C.F.R. 1926.652(b). See, *e.g.*, Greene Construction Co. Inc. 1 OSCH 1494 (1974); Underpinning & Foundation and Horn, 1 OSCH 3340 (1974).

49. 26 U.S.C.A. § 103(c)(4)(e).

50. Energy Supply and Environmental Coordination Act of 1974, Pub. L. No. 93-319 (1974), adding § 119 Clean Air Act. See Clean Air Act, § 119(k)(1)(E), 42 U.S.C.A. § 1857c-10.

51. Council of State Governments, "A Model Solid Waste Management and Resource Recovery Incentives Act" (1972) in *1973 Suggested State Legislation* 12.

52. *Id.,* § 4(9).

53. *Id.,* § 8.

54. *Id.,* §§ 4(14), 8(2)(3).

55. *Id.,* § 8(4)(5).

56. Council of State Governments, *1973 Suggested State Legislation* XII (1972).

57. COSGO Task Force Report, *supra* note 25 at 2. The District of Columbia also had a fairly comprehensive and unified administrative structure prior to the 1965 act.

58. *Id.* at 10. Local governments in the United States, whether urban or rural, are the legal creatures of the states. A.W. Bromage, "Local Governments," in W.B. Graves, *State Constitutional Revision* 240 (1960). For an example of a state constitutional provision including solid waste collection and disposal in the general police power authorization, see *Ga. Code Ann.* § 2-7901a(2)(1973).

59. But see COSGO Task Force Report, *supra* note 25 at 11, noting that at times local governments, on the basis of a narrow interpretation of their powers, considered solid waste collection and disposal services outside the welfare function granted to them and thus ignored these needs.

60. *Minn. Stat. Ann.* § 443.016 (1947) (repealed 1949).

61. *E.g., Ala. Code* tit. 37, § 496 (1959); *Mich. Comp. Laws* §§ 46.171 (1967) (counties), 123.301 (1967) (municipalities), 41.722 (1967) (towns); *Mo. Rev. Stat.* § 73.120 (1952).

62. COSGO Task Force Report, *supra* note 25 at 11, indicates that effective collection systems can be operated by local units with 10,000 population or more.

63. *Cf. Ga. Code Ann.* § 2-6001 (1973) (by local Amendment, proposed by Acts, 1918, p. 915, ratified Nov. 5, 1918).

64. *Ind. Ann. Stat.* § 19-2-15-2 (1974).

65. *Wis. Stat. Ann.* § 59.07(52) (1957).

66. *E.g., Ala. Code* tit. 37, § 496 (1959); *Kan. Gen. Stat. Ann.* § 12-2102 (1964).

67. *Mich. Comp. Laws* § 123.245 (1967).

68. *E.g., Del. Code Ann.* § 9-1524(b) (1974); *Ill. Rev. Stat.* ch. 34, § 5402 (1960); *Wash. Rev. Code* § 81.77.010 (Supp. 1972) (Regulatory Commission approval required).

69. *E.g., Ill. Rev. Stat.*, ch. 34, § 5402 (1960); *La. Rev. Stat.* §§ 33:401(8) (Supp. 1962) (municipalities), 33:1236 (Supp. 1962) (counties).

70. *E.g., Conn. Gen. Stat. Ann.*; § 7-161 (1972) (transport of refuse); *Del. Code Ann.* tit. 16, § 1708 (1974) (storage of refuse in multiple family dwellings).

71. *E.g., Del. Code Ann.* tit. 16, § 1701 (1974) (permit required for out-of-state refuse to enter).

72. COSGO Task Force Report, *supra* note 25 at 17.

73. *Minn. Stat. Ann.* § 368.79 (1966).

74. *Kan. Gen. Stat. Ann.* § 13-454 (1964).

75. *E.g., Ala. Code* tit. 37, § 496 (1959); *Ky. Rev. Stat.* § 94.285 (1969).

76. *E.g., Cal. Gov't. Code* § 25827 (1968).

77. *E.g., Fla. Stat. Ann.* § 210,03(5) (1972) (cigarette tax).

78. *Colo. Rev. Stat.Ann.*, § 30-20-203 (1973).

79. Solid Waste Disposal Act § 207, 42 U.S.C.A. §§ 3254a(b)(1), (2) (1970).

80. While many states title a statute a "Solid Waste Management Act" or some variation thereof, the name alone means little. Some states include solid waste provisions in comprehensive environmental legislation. In some cases the titled statue contains all the major elements, in others it covers primarily the administrative framework, with substantive duties delineated in other legislation.

The first statute cited for each state is the solid waste management act or its equivalent. The parenthetical date refers to the original enactment in response to the 1965 Federal Act. Where statutes have been renumbered or their location redesignated, current code references are cited though the date refers to the original state act. Dates of amendments are noted by an "a" after the date.

Ala. Code tit. 22, §§ 346-358 (1969a), tit. 37, § 496; *Alaska Stat.* §§ 46.03.010 *et seq.* (1971), § 29.48.033; *Ariz. Rev. Stat. Ann.* §§ 36-101 *et seq.* (1973), § 9-499, 511, 1221 *et seq.*, § 11-771; *Ark. Stat. Ann.* §§ 82-2701 *et seq.* (1971), §§ 19-2346, 2347, *Ark. Const.* art. 16, § 1, amend. 13; *Cal. Gov't. Code* §§ 66700-66793 (1972a), §§ 4500 *et seq.*, 25827, 25828, 25830, 38790-1, 39501, 54309.1, 54725, 54800 *et seq.*, 65451, *Cal. Health & Safety Code* §§ 215, 4100, 4121, 4170 *et seq.*, 4200, 4250, 4475, 4485, 5470, 14930, 39297.4; *Colo. Rev. Stat. Ann.* §§ 30-20-101 *et seq.* (1971a), § 20-201-205; *Conn. Gen. Stat. Rev.* §§ 19-524a *et seq.*, (1971a), §§ 7-148, 161, 162, 326, 339a *et seq.*, § 19-312; *Del. Code Ann.* tit. 7, §§ 6001-6033 (1973), tit. 9, § 1524, §§ 2400 *et seq.*, §§ 2900 *et seq.*, 4701, 4118, 4135, tit. 16, §§ 1701 *et seq.*, 1801 *et seq.*; *D.C. Code Ann.* (no solid waste law) §§ 6-501, 502, 508; *Fla. Stat. Ann.* §§ 403.701 *et seq.* (1974), *Fla. Const.* art 7, § 14, § 160.02, § 210.03, § 317.761, § 386.04, § 386.05, § 585.59; *Ga. Code Ann.* §§ 43-1601 *et seq.* (1972a), § 2-5701, § 2-5703, § 2-6001; *Hawaii Rev. Laws* §§ 342-1 *et seq.* (1972a), §§ 360-, 32-35, §§ 64-91, 95; *Idaho Code Ann.* §§ 31-4401 *et seq.* (1970a), § 39-103; *Ill. Rev. Stat.* ch. 111-1/2, §§ 1001 *et seq.* (1970), ch. 24, §§ 11-19-1 *et seq.*, 11-20-13, 11-80-10, ch. 34 §§ 417 *et seq.*, 5401 *et seq.*, ch. 139, § 39.15; *Ind. Ann. Stat.* §§ 19-2-1-1 *et seq.* (1965a), § 13-7-1-2, §§ 18-1-6-8, 15, § 18-1-18-1, § 18-5-1.5-3, §§ 19-2-6-1, 14, §§ 19-2-15-1 *et seq.*; *Iowa Code* §§ 455B.75 *et seq.* (1971a), § 135.11(7), § 136.3(2c), § 137.7, § 368.24, § 368.9(2,3), § 394.5, § 404.5(16), § 416.120; *Kan. Gen. Stat. Ann.* §§ 65-3401 *et seq.* (1973a), §§ 12-2101 *et seq.*, §§ 19-2658 *et seq.*, 2662, 2676, 2677, 3701, §§ 47-1301 *et seq.*, § 65-204, §§ 80-2201 *et seq.*; *Ky. Rev. Stat.* §§ 224.855 *et seq.* (1968a), §§ 94.281-287, §§ 109.010 *et seq.*; *La. Rev. Stat.* (no solid waste law), § 40:11.1, §§ 33:401, 1236, 1321 *et seq.*, 2922, 4161; *Me. Rev. Stat. Ann.* tit. 38, §§ 1301 *et seq.* (1973a), tit. 17, §§ 2251, 2253, tit. 30, §§ 417-22, 1202; *Md. Ann. Code* art. 43:387c *et seq.* (1965a), art. 23B, § 22, art. 25, §§ 3, 14A, *Natural Resources* §§ 3-101 *et seq.*, *Mass. Gen. Laws* ch. 16, §§ 1 *et seq.* (1969a), ch. 40: §§ 4, 44A-44K, ch. 40D: § 21; *Mich. Comp. Laws*

§§ 325.291 *et seq.* (1965a), §§ 41.721 *et seq.*, §§ 46.171 *et seq.*, §§ 123.241 *et seq.*, §§ 123.361 *et seq.*, §§ 123.731 *et seq.*; *Minn. Stat. Ann.* §§ 400.01 *et seq.* (1971a), § 115.01, §§ 116.01 *et seq.*, § 169.831, § 365.10, § 368.79, § 412.221, §§ 443. *et seq.*, § 460.58, § 473B.04, §§ 473D.01 *et seq.*; *Miss. Code Ann.* §§ 17-17-1 *et seq.* (1974), § 17-1-33, § 17-13-7, § 19-4-7, §§ 19-5-17 *et seq.*, §§ 19-5-151 *et seq.*, § 21-19-1; *Mo. Rev. Stat.* §§ 260.200 *et seq.* (1972), § 49.303, §§ 64.460 *et seq.*, §§ 71.680, 690, § 73.1205, § 74.137, § 251.180; *Mont. Rev. Codes Ann.* § 69-6001 *et seq.* (1965a), § 16-1031, §§ 46-2602 *et seq.*, §§ 69-4001 *et seq.*, 6001 *et seq. Neb. Rev. Stat.* §§ 81-1501 *et seq.* (1971a), §§ 19-2101 *et seq.*; *Nev. Rev. Stat.* §§ 444.440 *et seq.* (1971), § 244.187, §§ 268.081, .450, .712; *N.H. Rev. Stat. Ann.* §§ 147:23 *et seq.* (1967a), § 31:4, § 47:17, § 53-A:3, § 53-B *et seq.*, § 245:7; *N.J. Rev. Stat.* §§ 13:1E-1 *et seq.* (1970a), §§ 26:3-31, §§ 40:8B-1 *et seq.*, 14B-70, 37A-44 *et seq.*, 37C-1 *et seq.*, 63-43, 45, 51, 66-1 *et seq.*, 66A-31, 90-1 *et seq.*, 162A-1, 175-10, 23, §§ 48:13A-1 *et seq.*, §§ 52:34-21 *et seq.*; *N.M. Stat. Ann.* §§ 12-12-1 *et seq.* (1971a), §§ 14-49-1 *et seq.*, §§ 15-52-1 *et seq.*, §§ 15-57-1 *et seq.*; *N.Y. Env. Conserv.* §§ 27-0101 *et seq.* (1972a), § 3-0301(11), §§ 17-0303(5F), 0505, §§ 27-0501 *et seq.*, §§ 51-0901 *et seq.*, *County* § 226-B, §§ 250-276, *Gen. City* § 17, *Munic.* § 99-A, § 120-V, W, § 462(1A), *Pub. Auth.* §§ 1280 *et seq.*, *Pub. Health* §§ 1360-1364, §§ 1370-1373, *Second Class Cities* § 79, *Town* §§ 198(9), (9a), § 221, *Village* § 4-412(3), (4); *N.C. Gen. Stat.* §§ 130-166.16 *et seq.* (1969a), §§ 113A-1 *et seq.*, § 136-18.3, § 153-274, §§ 153A-132.1, 136, 292 *et seq.*, §§ 160A-192, 311 *et seq.*; *N.D. Cent. Code* §§ 23-29-01 *et seq.* (1975), § 11-11-14, § 11-28.1, §§ 40-22-01 *et seq.*, §§ 40-34-01 *et seq.*, § 61-02-21; *Ohio Rev. Code Ann.* § 3734.01 *et seq.* (1967a), §§ 343.01 *et seq.*, §§ 505.27 *et seq.*, § 715.43, § 717.01, § 735.02, § 3701.19, §§ 7123.01 *et seq.*; *Okla. Stat. Ann.* tit. 63, § 2251 *et seq.* (1970), tit. 63, §§ 1-908, tit. 82, §§ 1301 *et seq.*; *Ore. Rev. Stat.* §§ 459.005 *et seq.* (1967a), §§ 268.030, .310-.330, .350, § 450.075, §§ 451.010, .570; *Pa. Stat* tit. 35, §§ 6001 *et seq.* (1968a), tit. 3, §§ 452.1-452.15, tit. 16, §§ 1975, 2396, tit. 35, §§ 755.1 *et seq.*, tit. 52, §§ 30.51 *et seq.* 1 tit. 53, §§ 306, 2231-2234, 12675, 23127, 24691, 37403(6), (16), 46202(45), 56527, 65708, tit. 62 § 308; *R.I. Gen. Laws Ann.* § 23-46.1-1 *et seq.* (1968a), § 11-30-12, §§ 45-34-1 *et seq.*, *R.I. Solid Waste Corporation Act* ch. 176 PL1974 §§ 1-27; *S.C. Code Ann.* §§ 63-195 *et seq.* (1970), §§ 32-1275 *et seq.*; *S.D. Code* §§ 34-16B-1 *et seq.* (1972a), §§ 7-33-1 *et seq.*, §§ 9-32-10, 11; *Tenn. Code Ann.* §§ 53-4301 *et seq.* (1969a), §§ 5-1601 *et seq.*, 1901 *et seq.*, §§ 6-1901(1A), 2608, 2801, §§ 53-4502 *et seq.*; *Tex. Rev. Civ. Stat.* art. 447-7 *et seq.* (1969a), art. 969b, art. 976c, art. 4477-1, art. 6656; *Utah Code Ann.* (no solid waste law), § 10-8-61, § 10-11-1, §§ 17-29-1 *et seq.*, § 26-6-2, § 26-15-5; *Vt. Stat. Ann.* tit. 24, §§ 2201 *et seq.* (1967a); *Va. Code Ann.* § 32-9.1 (1970a), §§ 9-144 *et seq.*, §§ 15.1-28.1, 879; *Wash. Rev. Code* §§ 70.95.010 *et seq.* (1969), § 35.13.255, §§ 35.21.120 *et* seq., § 35.92.020, §§ 36.58A.010 et seq., §§ 36.82.240 et seq., § 70.95.090, § 81.77.015; W. Va. Code Ann. (via amendments to the following titles) §§ 7-1-3E, 5, 6, § 8-11-3, § 8-12-5, § 8-13-13, § 8-16-2, § 8-23-2, 3, 7, § 16-1-9, § 16-3-6, §§ 16-12-1 et seq., §§ 16-13-1 et seq.; Wis. Stat. Ann. §§ 499 et seq. (1967a), § 59.07, § 66.052, § 66.067, § 66.60, § 67.04, § 71.05, § 107.07, §§ 144.045 et seq., § 280.07; Wyo. Stat. Ann. §§ 35-462 et seq. (1973a), § 15.1-3, §§ 35-502.1, 35.502.56.

81. Besides Louisiana and Utah, Mississippi, North Dakota, and apparently South Carolina also issued rules and regulations under preexisting state health department authority before actually passing new legislation on solid waste.

82. Letter from Charles Porter, supervisor, Solid Waste Division of Wyoming Department of Environmental Quality, to Daniel C. Goldfarb, June 2, 1975, on file at Legislative Drafting Research Fund of Columbia University.

83. See COSGO Task Force Report, *supra* note 25 at 16, 19.

84. The transfer of solid waste management jurisdiction from public health to environmental protection agencies occurred in Illinois (1970), Kentucky (1968), Michigan (1973), New Jersey (1970), New Mexico (1973), Ohio (1974), South Dakota (1974).

85. Ala., Ariz., Colo., Fla., Hawaii, Idaho, Ind., Kan., La., Md., Mass., Miss., Mont., Nev., N.H., N.C., N.D., Okla., R.I., S.C., Tenn., Tex, Utah, Va., W.Va., Wyo. Of these 26, 10 are designated as environmental control bureaus, or the equivalent, within the state health department.

86. Alas., Ark., Cal., Conn., Del., D.C., Ga., Ill., Iowa, Ky., Me., Mich., Minn., Mo., Neb., N.J., N.M., N.Y., Ohio, Ore., Pa., S.D., Vt., Wash., Wis.

87. COSGO Task Force Report, *supra* note 25 at 31.

88. One of the goals of the federal legislation is to produce uniformity among state laws, Solid Waste Disposal Act § 206, 42 U.S.C. §3254 (1970), and, quite understandably, federal legislation has been achieving this goal much faster in the area in which it has concentrated, disposal, than in collection, which is a secondary concern.

89. It should be noted that the extent of regulation varies greatly, and that some states have very limited regulatory provisions, *e.g.*, Md. and Ky. Full citations appear in note 80 *supra*.

90. The seven states lacking express statutory authority to issue guidelines on collection are: Colorado, Montana, New Hampshire, Ohio, Tennessee, Virginia, and West Virginia.

91. *Ohio Rev. Code Ann.* § 3734.02(D) (1974).

92. Del., Fla., Ill., Ind., Ky., Me., Md., Mass., Mo., Neb., N.Y., R.I., and Vt.

93. Ala., Alas., Ariz., Ark., Cal., D.C., Ga., Hawaii, Idaho, Iowa, Kan., La., Mich., Minn., Miss., Nev., N.J., N.M., N.C., N.D., Okla., Ore., Pa., S.C., S.D., Tex., Wash., Wis. The regulations issued by the state health department in

Louisiana, one of the two states acting under preexisting authority and not under recent solid waste management legislation, does cover collection as well as disposal, though Utah's is for disposal only.

94. Hawaii Pub. Health Dept. Reg., Ch. 46, "Solid Waste Management Control," § 6E (1974).

There is, of course, no clear-cut distinction between statutes and regulations in terms of where standards might be included; generally, however, powers and duties relating to solid waste management are included in the statutes, whereas specific standards are left to the regulations.

95. Solid Waste Disposal Act § 208, 42 U.S.C. § 3254b(c) (1970).

96. Alas., Ariz., Ark., Cal., Conn., Del., Fla., Kan., Mass., Neb., N.J., N.Y., N.C., N.D. (disposal only), Okla., Pa., R.I., Tex., Va., Wash.

97. By June 1, 1972, forty-six states had solid waste management plans completed or in draft stage, and the remaining four were in inventory stage. Council of Environmental Quality, *Environmental Quality-Third Annual Report*, p. 176 (1972).

98. Ariz., Ark., Cal., Conn., Del., Fla., Ga., Kan., Md., Mich., Minn., Mo., Nev., N.Y., N.D., Pa., R.I., S.D., Va., Wash.

99. Ga. Dept. of Natural Resources, "Rules and Regs. for Solid Waste Management," ch. 391-3-4.05(2)(a) (1974).

100. *Kan. Stat. Ann.* §§ 65-3402, 3405 (Supp. 1974).

101. See section above, "The Growing Role of State Governments: The Early Phase."

102. Ala., Alas., Ariz., Ark., Cal., Colo., Conn., D.C., Ga., Idaho, Ill., Ind., Ky., La., Me., Md., Mich., Miss., Mont., Neb., Nev., N.H., N.J., N.M., N.Y., N.C., Ohio, Okla., S.C., Tenn., Tex., Utah, Va., Wash., W.Va., Wis.

103. Del., Fla., Ind. (for ashes and putrescible wastes in first-class cities), Iowa, Kan., Minn., Mo., N.M., N.D., Pa., S.D., Wis. (Milwaukee County).

104. *E.g., Mo. Rev. Stat.* § 260.215 (Supp. 1975).

105. *E.g.,* Iowa Dept. of Environmental Quality, "Rules and Regs. Relating to Solid Waste Disposal" § 26.5(1) (1971).

106. N.D. Dept. of Health Reg. No. 86, "Solid Waste Management Regs." § 4.1 (1970).

107. Minn. Division of Solid Waste, Solid Waste Disposal Regs. § 11(3) (1970).

108. *Pa. Stat.* tit. 35 § 6005(a) (Supp. 1975). Another way in which this distinction is recognized is in different deadlines for implementation. *E.g.,* S.D. grants counties and municipalities with populations under 10,000 two years longer to implement solid waste management systems than it does those with populations over 10,000. S.D. Dept. of Environmental Protection, Rules. Ch. 34:17:01:04-05 (1974).

109. Kan., N.M., N.D., Pa., S.D.

110. Ariz., Ark., Cal., D.C., S.C., Wash.

111. Ariz. Environmental Health Services, Solid Waste Rules § 2-4-3.1 (1973); D.C. Reg. No. 71-21, "Solid Waste Disposal Regulations" 8-3:604(b) (1971); S.C. State Board of Health, SBH-SW Reg. 2 § IV(E) (1972).

112. S.C. State Board of Health, SBH-SW Reg. 2 § IV(E) (1972).

113. Ark. Dept. of Pollution Control and Ecology, "Ark. Solid Waste Disposal Code" § 12(d) (1973); Pa. Dept. of Environmental Resources, Rules and Regs., ch. 75, "Solid Waste Management" § 75.164 (1971).

114. See ch. 6 *supra.*

115. Ala. Alas., Ariz., Ark., Cal., Colo., Conn., Del., D.C., Fla., Idaho, Ill., Ind., Iowa, Kan., Ky., Me., Mich., Minn., Miss., Mo., Mont., Neb., Nev., N.J., N.M., N.Y., N.C., N.D., Okla., Pa., S.C., S.D., Tenn., Tex., Va., Wash., W.Va., Wis. (Milwaukee County).

116. Ala., Alas., Ariz., Ark., Cal., Colo., Conn., Fla., Ga., Idaho, Ill., Ind., Iowa, Kan., Ky., Me., Md., Mass., Mich., Minn., Miss., Mo., Neb., N.H., N.J., N.Y., N.C., N.D., Okla., R.I., S.C., S.D., Tenn., Tex., Vt., Va., Wash., W.Va., Wis.

117. *Kan. Gen. Stat. Ann.* § 19-2677 (1974); *Ore. Rev. Stat.* §§ 459.065, 459.085 (1973); *S.D. Code* §§ 34-16B-17.1, 34-16B-21.1 (Supp. 1974).

118. *Kan. Gen. Stat. Ann.* § 19-2677 (1974); *S.D. Code* §§ 34-16B-17.1, 34-16B-21.1 (Supp. 1974).

119. *Mo. Rev. Stat.* § 73.125 (Supp. 1975).

120. *N.C. Gen. Stat.* § 153A-136(a)(3) (1974).

121. *Mich. Comp. Laws* § 123.245 (1967).

122. *N.Y. Munic.* § 120W (Supp. 1974).

123. *Ariz. Rev. Stat. Ann.* §§ 9-1221 *et seq.* (Supp. 1974).

124. *Conn. Gen. Stat. Ann.* § 19-524 ff. (Supp. 1975); *R.I. Gen. Laws Ann.* § 23-46-3 (Supp. 1974); *S.D. Code* § 34-16B-23.3 (Supp. 1974).

125. *Cal. Gov't. Code* § 66771 (Supp. 1975).

126. Cal. Solid Waste Management Board, Regulations §§ 17331 *et seq.* (1974). The idea that solid waste collection is a matter of statewide concern is by no means universally accepted yet. As noted, *supra* note 12, Ohio's statute excludes collection from the state agency's jurisdiction, and the Mass. Solid Waste Management Plan at page 33 speaks of "the collection and haul of refuse as a local responsibility."

127. Ala., Alas., Ariz., Ark., Cal., Fla., Ga., Hawaii (on matters as yet unregulated by the state), Ind., Kan., La., Me., Md., Mich., Miss., Mo., Mont., Neb., Nev., N.H., N.J., N.M., N.Y., N.C., N.D., Okla., Ore., Pa., S.C., S.D., Tenn., Tex., Wash., Wis.

128. *Hawaii Rev. Stat.* § 342-19(b) (Supp. 1974).

129. Ala., Ariz., Ark., Cal., Conn., Del., Hawaii, Idaho, Ill., Ind., Kan., Ky., La., Md., Minn., Miss., Mo., Nev., N.J., N.M., N.C., N.D., Okla., Ore., Pa., S.C., S.D., Tenn., Tex., Wash., Wis.

130. Ala., Ariz., Ark., Cal., Conn., Iowa, Kan., Ky., La., Miss., Neb., Nev., N.J., N.M., N.C., N.D., Okla., Ore., Pa., R.I., S.D., Tenn., Utah, Wash., W.Va., Wyo.

131. See note 22 *supra.*

132. See note 29 *supra.*

133. Ala., Alas., Ariz., Ark., Cal., Del., D.C., Fla., Hawaii, Idaho, Ind. (regulated by fire marshal), Iowa, Kan., La., Me., Minn., Miss., Neb., Nev., N.M., N.C., N.D., Ore., Pa., S.C., S.D., Wash., Wis.

134. Ala., Alas., Ariz., Ark., Cal., Colo., Conn., Del., D.C., Fla., Ga., Hawaii, Kan., La., Mich., Minn., Miss., Neb., Nev., N.M., N.C., N.D., Okla., Ore., Pa., S.C., Tex., Utah, Wash., Wis.

135. Idaho Dept. of Environmental and Commty. Services, "Solid Waste Management" Regs. and Standards" § 5.1 (1973). See also Hawaii Regs., *supra* note 15.

136. Ala. Board of Health, "Rules and Regs. for Solid Waste Management" § V(b) (1972).

137. Kansas Dept. of Health, "Solid Waste Management Standards and Regs." § 28-29-8(B) (1972). Kansas's regulations on storage are as detailed as any and seek to provide convenience and minimum unsightliness and odors, as well as the usual public health assurances.

138. *E.g.,* N.C. State Board of Health, "Rules and Regulations Providing Standards for Solid Waste Disposal," RIVB(2) (1971).

139. *Ark. Stat. Ann.* §§ 82-2705(e) (municipalities), 2706(e) (counties), (Supp. 1973).

140. *Conn. Gen. Stat. Ann.* § 19-524r(10) (Supp. 1975).

141. Ala., Alas., Ariz., Cal., Conn., Fla., Ga., Hawaii, Idaho, Ill., Iowa, Kan., Me., Md., Mich., Minn., Miss., N.H., N.J., N.Y., N.D., Ore., Pa., R.I., S.C., S.D., Tenn., Tex., Vt., Wash., Wis.

142. *E.g.,* Kansas Dept of Health, "Solid Waste Management Standards and Regs." § 28-29-9(B) (1972).

143. *E.g.,* Mich. Dept. of Natural Resources, Solid Waste Management Division, "Rules Governing Solid Waste Management" § 325.2745(1) (1973).

144. *E.g.,* Ga. Dept. of Natural Resources, "Rules and Regs. for Solid Waste Management," ch. 391-3-4-.06(3) (1974).

145. *E.g.,* N.C. Board of Health, "Rules and Regs. Providing Standards for Solid Waste Disposal" § V(B) (1971).

146. *E.g.,* N.D. State Dept. of Health, "Solid Waste Management Regs." § 4.3 (1970).

147. *E.g.*, Mich. Dept. of Natural Resources, Solid Waste Management Division, "Rules Governing Solid Waste Management" § 325.2745(3) (1973).

148. Ala., Cal., Del., D.C., Fla., Ga., Ill., Ind., Kan., Md., Mass., Mich., Minn., Miss., Nev., N.J., N.M., N.C., N.D., S.C., Wis.

149. *E.g., Ill. Rev. Stat.* ch. 111-1/2 § 1021 (Supp. 1975).

150. *E.g.,* Ga. Dept. of Natural Resources, "Rules and Regs. for Solid Waste Management," ch. 391-3-4-.02(3) (1974).

151. *E.g.,* Cal. Solid Waste Management Board, Regs. § 17332 (1974).

152. Mich. Dept. of Natural Resources, "Rules Governing Solid Waste Management" R325.2743 (1975); Miss. State Board of Health, "Solid Waste Management Regs." § VII.B.4 (1974).

153. See section above, "Other Federal Legislation."

154. *Minn. Stat. Ann.* § 169.831 (Supp. 1975).

155. *Miss. Code Ann.* § 63-5-51 (1973).

Appendix 11A
State Legislation and Regulation for Solid Waste Management

Table 11A-1, indicating the powers and duties of state and local governments for solid waste management, is based on the solid waste management legislation and regulations of the fifty states and the District of Columbia, as well as on other state statutes which refer specifically to aspects of solid waste management. An "a" after a year, for example, 69a, means that the enactment of 1969 has been significantly amended since. "S" refers to statutes and "R" to rules and regulations propulgated by state agencies.

The mark "x" in column means that a state has an agency or rules and regulations governing solid waste. Thus, for example, under Roman numeral I—Collection—"state agency authorized to regulate," an "x" under a particular state indicates that the state has such an agency.

A blank on the chart does not indicate that the particular function is not carried out or that the particular exercise of authority is denied or prohibited. It simply means that the statutes and regulations of the state have not referred to the particular matter specifically. The chart indicates that the statutes and regulations of the fifty states do not themselves mandate the patterns or arrangements for solid waste management which were found through empirical survey and are indicated in other parts of the study, and that they do not *by themselves* explain such patterns or arrangements.

	Al	Ak	Az	Ar	Ca	Co	Ct	De	DC	Fl	Ga	Hi	Id	Il	In	Ia	Ks	Ky	La	Me
Adoption of major SW legislation	69a	71	73	71	72a	71	71a	66a		74	72a	72a	70a	70	65a	71a	71	66a		73a
Adoption of major SW regulations	72	73	73	73	74	72	75	68a	71	71a	72a	74	73	66a	68a	71a	72	68a	68	74
I. COLLECTION																				
State agency authorized to regulate	X	X	X	X	X		X	X	X	X	X	X	X	X	X	X	X	X	X	X
Regulations have been issued	X	X	X	X	X				X		X	X	X			X	X		X	
Counties: must arrange for collection								S		S						R	R			
authorized to arrange for collection	S		S	S	S	S					S		S		S			S	S	
authorized to contract out for services	S		S	S	S	S		R		S			S		S	S	S	S		S
authorized to regulate collection	S		S	S	S	S		S				S	S	S				S	S	
Municipalities: must arrange for collection										S						R	R			
authorized to arrange for collection	S	S	S	S			S		S					S	S			S	S	S
authorized to contract out for services	S	S	S	S	S		S	R	S	S				S	S	S	S	S		S
authorized to regulate collection	S	S	S	S	S		S									S	S	S	S	
Licenses or permits required for collectors	S				R			S	R	S	S			S	R		R			
Storage of refuse regulated	R	R	R	R	R			S	R	R		R	R			S	R	R		R	R
Transport of refuse regulated	S	R	R	R	R	S		S	R	S	R	R					R		R	
Minimum frequency of pickup prescribed			2/wk	1/wk	1/wk			2/wk								1/wk				
II. GENERAL STATUTORY PROVISIONS																				
SW management plan required from:																				
state agency		S	S	S	S		S	S		S							S			
counties			S	S	S			S		S	S						S			
municipalities			S	S	S		S			S	S						S			
Stricter local laws permitted	R	R	S	S	S					S	S	S			S		R		S	S
Intergovernmental cooperation authorized	S	S	S	S	S	S	S			S	S		S	S	S	S	S	S		S
State authorized to:																				
provide technical assistance		S	S	S	S	S				S	S	S			S		S			S
provide financial assistance		S	S		S		S	S		S	S						S			S
conduct research programs		S	S	S	S			S		S	S	S		S			S			
Relationship indicated to:																				
air pollution control	S	S		S	S	S	S	S	R	S	S	S	R	S	S	R	S	S		S
water pollution control	S	S	R	S	S	S	S	S		S	S	S	R	S	S	R	S	S	S	S
land use planning				R	S		S	S		R			S	S	S					
recycling & resource recovery	R	R	S		S		S			S	R	S	R	S		R	R			R
III. DISPOSAL																				
State agency authorized to regulate	X	X	X	X	X	X	X	X	X	X	X	X	X	X	X	X	X	X	X	X
Regulations have been issued	X	X	X	X	X	X	X	X	X	X	X	X	X	X	X	X	X	X	X	X
Counties: must arrange for disposal				S				R		S			S			S	R			
authorized to arrange for disposal	S		S		S	S					S			S	S			S	S	
authorized to contract out for services	S		S	S	S			R		S				S	S	S	S	S	S	
authorized to regulate disposal	S		S	S		S							S	S	S	S	S	S	S	
Municipalities: must arrange for disposal				S			S	R		S						S	R			S
authorized to arrange for disposal	S	S	S		S	S		S						S	S			S	S	
authorized to contract out for services	S	S	S	S	S	S	S	R	S	S				S	S	S	S	S	S	S
authorized to regulate disposal		S	S			S								S		S	S	S	S	
Licenses or permits required for disposal	S	S	S	R	R	S	S	S	R	S	S	S	R	S	R	S	S	S	R	S
Approval required for new disposal sites	R	R	S	R	R	S	S	R	R	S	S	S	R	R	S	S	S	S	R	R
Hazardous wastes regulated	S	S	R	R	S	S	S		R	R	R		R	R	S	S		S		R
IV. ENFORCEMENT																				
Right of inspection	S	S	R	S	R		S	S	R	R	S	S	S	S		S	S	R		
Power to revoke licenses or permits	S	R				S	R	R	R		S	S	R	R	S	S	S			
Administrative proceedings		R	S	S		R	S				S	S	S	S	S	S	S			
Injunctive powers			S	S	S		S		S	S	S	S	S		S	S	S			S
Civil penalties		S					S	S		S	S	S	S							S
Criminal penalties	S	S	S	S		S		S	S		S	S	S	S		S				

	Mi	Ma	Mi	Mn	Ms	Mo	Mt	Ne	Nv	NH	NJ	NM	NY	NC	ND	Oh	Ok	Or	Pa	RI	SC	SD	Tn	Tx	Ut	Vt	Va	Wa	WV	Wi	Wy
	65a	69a	65a	71a	74	72	65a	71a	71a	67a	70a	71a	66a	69a	75	67a	70	67a	68a	68a	70	72a	69a	69a		67a	70a	69	73	67a	73a
	70	71	65a	70	71a	73	66a	72	73	72	70a	74	73a	71	70	68a	71	72a	71a	75	71	74	71	70a	74	69	71	72	74	69	
	x	x	x	x		x		x	x		x	x	x	x	x	x	x	x	x		x	x		x	x	x		x		x	x
			x	x	x				x		x	x		x	x		x	x	x		x	x		x				x		x	
				R		s									R							s									
	s		s		s		s		s		s	s	s	s		s	s				s		s	s		s	s	s	s		
			s	s	s	s	s		s		s	s	s	s	R	s					s	s	s	s		s	s	s		s	
	s			s	s				s		s		s	s		s	s	s			s	s	s				s		s		
						s						R			R				s				s								
	s		s		s		s	s	s	s	s		s	s		s	s				s		s		s	s	s	s			
			s		s		s	s	s		s	s			R	s			s		s	s				s	s	s			
					s				s		s			s	s			s	s	s	s	s		s			s	s		s	s
	s	s	s	s	s				s		s	R		s	s						s									R	
				R	R				s	R		R		R	R			R	R		R	R			R	R		R		R	
			s	s	s				s	R		R		R	R		R	R	R		R			R	R		s		R		
												1/4k			1/4k				1/4k		2/4k	1/4k			1/4k						
		s									s		s	s		s			s	s				s			s	s			
	s		s	s		s			s				s		s				s	s		s		s			s	s			
	s		s	s		s			s				s		s				s	s		s		s			s				
	s		s	s	R	s	s	s	s	s	s	s	s	s	s	s		s	s		R	s	s	R	s			s		s	
	s		s	s	s	s		s		s	s	s	s	s	s	s				s	s	s	s	s		s		s		s	
	s	s	s			s		s		s	s		s		s	s		s	s	s	s	s	s		s		s	s		s	s
	s		s			s		s					s	s					s			s	s		s			s			
	s	s	s					s			s		s						s	s	s					s		s		s	
	s	R	R	s	s	s	s	s	R			s	s	R	s			s	s	R	R	s	s	s	R	R	R	s	R	R	s
	s	R	R	s	s	s	s	s	R		s	s	s	R	s	s	R	s	s	R	R	s	s	s	R	R	s	R		R	s
	s		R	s	R	s		s			s		s					s	s	s	s		s	s		s	s		s	R	s
	s		R	s	R					s	s		s		s			s	s	s	R	s	s	s		s	R		s		
	x	x	x	x		x	x	x	x	x	x	x	x	x	x	x	x	x	x	x	x	x	x	x	x	x	x	x	x		x
	x	x	x	x	x	x	x	x	x	x	x	x	x	x	x	x	x	x	x	x	x	x	x	x	x	x	x	x	x		
				R		s			s		R											s					s				
	s	s	s		s		s				s		s	s	s	s	s	s	s		s		s	s			s	s			
	s	s		s	s	s	s		s		s	s	s	s				s			s	s	s	s		s					
			s	s	s	s			s		s		s		s			s	s		s	s	s	s				s		s	
		s			s	s		s	s	s			R						s	s			s			s					
	s		s	s	s	s	s		s		s		s	s	s	s		s	s	s	s	s	s			s		s	s		s
	s		s			s	s				s		s	s		s	s				s		s			s		s			
	s			s		s			s			s		s	s	s	s		s	s	s		s		s		s	s	s	s	s
	s	s	s	s	s	s	s		s		s	R	s	s	s	s		s	s	s	s	s	s	s	R	R	s	s	s	s	s
	s	s	s	s	s	s	s	s	R	s	s	R	s	R		s		s	s	s	s	s	s	s	R	R	s	s	s	s	s
		R	s	s	s	R	R	R	R	s	s		s	R	R		s	s	R			R	R	R	R	R	R	R		R	
	s	R	s			s	s	s	s				s		s	s	s	s	s	R	s	s	s	s	s	s	R		s	s	
	s	s				s	s					s	R	s		s	s	s	s		R	s	s	s	s		R	s	E	R	s
	s	s				s		s					s			s	s	s	s		s		s	s	s			s	E	s	s
			s	s		s		s	s	s	s		s			s		s	s		s	s	s							s	
	s	s				s	s	s	s	s	s		s	s	s	s		s	s	s	s		s	s	s				s	s	s

12 Policy Conclusions and Recommendations

This study has been completed at a time when the performance of government is being scrutinized and its ability to deliver public services efficiently is being questioned. The findings reported indicate that the questioning is not misplaced, and that at least with respect to one essential municipal service there is an efficient alternative to a city agency. It is indeed possible to organize and deliver an important public service through the private sector while retaining adequate public control. *Cities larger than 50,000 population are able to purchase residential refuse collection services from private firms at a lower cost than they can provide it themselves through their own organization and work force.*[1]

Cities that are under pressure to reduce expenditures would be well advised to look closely at their solid waste collection practices. It is easy enough to reduce costs by reducing the level of service—for example, by reducing the frequency of service or by switching from backyard to curbside pickup—but a reduction in citizen service cannot properly be called an increase in efficiency. The challenge facing local government officials is to reduce costs while maintaining the level and quality of service.

One possible area for improvement is in the management practices identified in Chapter 8. The greater efficiency of contract collection, compared to municipal collection, seems attributable at least in part to the larger crews, smaller trucks, more rear-loading trucks, and higher absenteeism found in municipal agencies. If these practices were changed so that city agencies resembled private firms more closely in these terms, then the cost of municipal collection might be driven down to more nearly match the lower cost of contract collection. This is certainly a direction to explore, but it should be noted that in many cities it is very difficult to implement these changes, and the best way to do so may well be to change to contract collection, or, at the very least, to threaten to do so.

While the findings here refer to cities larger than 50,000 in population, the effect of taxes should not be overlooked. As was pointed out in Chapter 8, when citizens pay the government $100 for refuse collection services, they receive only those services, but when they pay an equal amount to a private firm they receive not only the service but also a rebate, so to speak, of $15 worth of other government services by virtue of the fact that the firm will pay about $15 in taxes out of its gross income of $100. This means that in cities smaller than 50,000, where the difference in calculated cost between municipal and contract collection is not significant, the 15 percent "tax rebate" is significant and results in a better bargain for the citizen. In other words, citizens in any size community will have more net services per dollar through contract collection than through municipal or private collection.

An important caveat to be borne in mind is that only efficiency was studied, and it was measured narrowly as the cost to the household for collection service (excluding disposal costs). The *effectiveness* of the different arrangements was not studied and one might argue that contract collection achieves efficiency by sacrificing quality, reliability, responsiveness, courtesy, promptness, cleanliness, and so forth. While this issue has not yet been studied in detail, a survey of households receiving private collection indicates a very high degree of satisfaction with their service and, although this is by no means conclusive evidence on the subject, there is no evidence to support the argument that the private sector attains efficiency at the expense of effectiveness.

It should also be clearly understood that the finding concerning contract and municipal collection must not be misinterpreted to mean that all cities larger than 50,000 would lower their costs if they were to change from municipal to contract collection. As the data show, there is considerable overlap and there are many large cities with efficient municipal collection and many others with inefficient contract collection. Indeed, some cities with contract collection would be well advised to change to municipal collection as a means of reducing cost.

Nevertheless, the general finding holds true, and the author believes it proper to interpret the findings in this way: if one were to choose at random a hundred large cities (greater than 50,000 population) with municipal collection, and each were to switch to contract collection, the total cost of collection for the hundred cities taken together would decline, although the cost would probably increase for some of the individual cities.

What then should a cost-conscious city official do? The starting point is to determine the current total cost of refuse collection in the community, regardless of arrangement. This is relatively easy to do for contract and franchise collection, but one must remember to add the city's cost of monitoring and supervision. It takes some effort to determine costs if the city has private collection. It generally requires the greatest care and determination to ascertain the cost of municipal collection: cities basically do not know this figure, for their budgets rarely reveal the full cost. Evidence presented in Chapter 6 indeed demonstrates that cities do not know how much municipal collection costs them. Striking confirmation of this assertion was obtained by comparing budgeted costs with actual costs in the sample of cities with municipal collection that were visited. The average ratio of actual costs to budgeted costs for sixty-eight cities was 1.30; that is, the actual cost was 30 percent greater than the cost shown in the budget for that service.

Budgeting and accounting practices differ greatly among cities, however the following is a useful composite list of the common reasons why budgets usually fail to display the full costs of municipal refuse collection:

1. Capital costs of vehicles do not always appear in the budget of the department using the vehicles.
2. The cost of fuel and oil and other operating costs of vehicles are also often missing from the budget of the refuse collection agency.
3. Labor costs for vehicle repairs often appear elsewhere in the budget, depending on how vehicle maintenance is carried out.
4. The cost of fringe benefits, including pension contributions, may or may not be allocated to the department where a worker is assigned.
5. Pension funds are seriously underfunded in many cities, so that the annual pension contribution paid by a city does not fully reflect the cost that will ultimately be incurred by the city.
6. The cost of supplementary workers borrowed from other departments to fill in during vacations and absences of regular workers is sometimes overlooked.
7. The costs of constructing, maintaining, and operating garages and offices may not be included in department budgets.
8. The cost of the property used by the agency, in the form of property taxes that would be collected if the property were privately owned, is rarely treated as a cost of the service.
9. Overhead costs are similarly ignored, in many cases.
10. The cost of insurance premiums (for example, for vehicle and personal liability) is often excluded from departmental budgets; if the city is self-insured, the cost of relevant liability claims paid should likewise be included in the cost of the service.

The cost of service having been determined, the findings in the various chapters should be considered for local applicability. A small city might lower its cost of service, regardless of arrangement, by forming a larger market. Giving exclusive service rights in an area to a single service provider, public or private, is likely to reduce costs. Paying for the service through taxes has some virtues, including a reduction of billing costs, better enforcement of mandatory collection ordinances, and federal income tax savings for certain taxpayers.

In particular, the potential savings of a change to contract collection (or mandatory franchise collection) should be evaluated locally, whether the city has municipal or private or nonmandatory franchise collection. This might best be done by soliciting formal or informal bids from private refuse collection firms, after thoroughly studying the city's current system.

The average city of over 50,000 population can reasonably expect to achieve significantly lower costs by contracting for service, provided that its procedures for awarding contracts are at least of average effectiveness and provided that the refuse collection industry in that location is at least as competitive as it is in the average community. (Chapter 10 provides useful insights as to contract terms and, although Chapter 9 points out that competitive bidding leads to prices no different than negotiated bidding, many cities will no doubt prefer the former, if only for the sake of appearances.) Smaller cities, on average, should experience no increase in cost if they were to change from municipal to contract collection, and could increase their revenues as a result of taxes paid by the contractors.

Cities with greater than 100,000 population have an interesting opportunity to introduce competition. Such cities can be divided into two or more districts, each with at least 50,000 population, and have different service providers in the different districts without foregoing any economies of scale. A municipal agency could service one or more of the districts, while the city could contract with one or more private firms to service the remaining district or districts. By doing so, a large city might best assure a continued competitive environment and protect itself against possible future collusion by its contractors or coercion by its employees. Such an arrangement might also be expected to reduce the city's risk of service disruptions owing to strikes or business failures.

A city's basic strategy should be to retain some options in service delivery and to have the continuing capability of measuring and comparing performance. It is to a city's advantage to create a healthy competitive climate that will lead to increased productivity and more cost-effective delivery of services to citizens. A mayor or a city manager should be in the position of a manufacturing executive who has the choice of manufacturing a part within his factory or of purchasing it from an outside supplier.

A long-term monopoly, whether public or private, is likely to be unsatisfactory, both in manufacturing and in refuse collection. Municipal collection comes close to being a permanent monopoly. On the other hand, continuous competition—that is, private collection—was shown to be inefficient. The system of temporary monopoly, or periodic competition, as exemplified by contract collection, seems to be best.

Note

1. The policy conclusions reached in this study are corroborated by two more limited research efforts that were concurrent with this work. In a study of cities in Connecticut, contract collection appeared to be less costly than municipal collection. See Peter Kemper and John M. Quigley, *The Economics of Refuse Collection* (Cambridge, Mass.: Ballinger, 1976). A study of cities in Switzerland provides international support for our findings, concluding that refuse collection by public agencies seems to be more costly than refuse collection by private firms. See Werner W. Pommerehne and Bruno S. Frey, "Public vs. Private Production Efficiency: A Theoretical and Empirical Comparison," in *Comparing Urban Service Systems: Relations Between Structure and Performance*, ed. Vincent Ostrom, Urban Affairs Annual Reviews, vol. 11 (Beverly Hills, Calif.: Sage Publications, 1977).

Appendixes

Appendix A
Universal Telephone Survey

Sample Selection

The purpose of the Universal Telephone Survey (UTS) was to provide basic information about solid waste collection practices. This information included service levels and service arrangements used by different types of service recipients.

Since this survey applies to the entire universe of cities eligible for this survey and since samples for all subsequent surveys were selected based on results of this survey, the sampling procedure used to select survey recipients is described in great detail. The survey instrument and survey methods are then described.

SMSA Identification

Of the 243 Standard Metropolitan Statistical Areas (SMSAs) defined by the 1970 census, only 200—those which met the following criteria—were eligible according to the terms of the solicitation issued by the National Science Foundation:

1. Location within one, and only one, of the fifty states;
2. Maximum SMSA population of 1,500,000 in the 1970 census.

Eligible SMSAs and Their Populations

The 200 SMSAs and their populations are grouped by order of quadrant, region, state, and alphabetically within the state. (See Table A-1.)

Eligible Local Jurisdictions

Organized local jurisdictions with populations exceeding 2,500 wholly or partially within the 200 SMSAs were identified as the study units. These include incorporated municipalities, towns, and townships.

Cities eligible for the collection study and thus the Universal Telephone Survey were: all incorporated municipalities above 2,500 population by the 1970 Census; and all towns and townships with state-provided legal authority to collect or arrange for collection of solid waste, provided that the portion of the

Table A-1
200 Eligible SMSAs by Region and State

Region	State	SMSA	Population
Northeast Quadrant			
New England	Connecticut	Bridgeport	389,153
		Bristol	65,808
		Danbury	78,405
		Hartford	663,891
		Meriden	55,959
		New Britain	145,269
		New Haven	355,538
		New London-Norwich	208,412
		Norwalk	120,099
		Stamford	206,419
		Waterbury	208,956
	Maine	Lewiston-Auburn	72,747
		Portland	141,625
	Massachusetts	Brockton	189,820
		Fitchburg-Leominster	97,164
		Lowell	212,860
		New Bedford	152,642
		Pittsfield	79,727
		Worcester	344,320
	New Hampshire	Manchester	108,461
		Nashua	66,548
	Rhode Island[a]		
	Vermont[a]		
Middle Atlantic	New Jersey	Atlantic City	175,043
		Jersey City	609,266
		Paterson-Clifton-Passaic	1,358,794
		Trenton	303,968
		Vineland-Bridgeton-Millville	121,374
	New York	Albany-Schenectady	721,910
		Buffalo	1,349,211
		Rochester	882,667
		Syracuse	636,507
		Utica-Rome	340,670
	Pennsylvania	Altoona	135,536
		Erie	263,654

Table A-1 (cont.)

Region	State	SMSA	Population
Middle Atlantic (cont.)		Harrisburg	410,626
		Johnstown	262,822
		Lancaster	319,693
		Reading	296,382
		Scranton	234,107
		Wilkes-Barre-Hazelton	342,301
		York	329,540
North Central Quadrant			
East North Central	Illinois	Bloomington-Normal	104,389
		Champaign-Urbana	163,281
		Decatur	125,010
		Peoria	341,979
		Rockford	272,063
		Springfield	161,335
	Indiana	Anderson	138,451
		Fort Wayne	230,455
		Gary-Hammond-E. Chicago	633,367
		Indianapolis	1,109,882
		Lafayette-W. Lafayette	109,378
		Muncie	129,219
		South Bend	280,031
		Terre Haute	175,143
	Michigan	Ann Arbor	243,103
		Bay City	117,339
		Flint	496,658
		Grand Rapids	539,225
		Jackson	143,274
		Kalamazoo	201,550
		Lansing	378,423
		Muskegon-Muskegon Hts.	157,426
		Saginaw	219,743
	Ohio	Akron	679,239
		Canton	372,210
		Columbus	915,228
		Dayton	850,266
		Hamilton-Middleton	226,207
		Lima	171,472
		Lorain-Elyria	256,843

Table A-1 (cont.)

Region	State	SMSA	Population
East North Central (cont.)		Mansfield	129,997
		Springfield	157,115
		Youngstown	536,003
	Wisconsin	Appleton-Oshkosh	276,891
		Green Bay	158,244
		Kenosha	117,917
		LaCrosse	80,468
		Madison	290,272
		Milwaukee	1,403,688
		Racine	170,838
West North Central	Iowa	Cedar Rapids	163,213
		Des Moines	286,101
		Dubuque	90,609
		Waterloo	132,916
	Kansas	Topeka	155,322
		Wichita	389,352
	Minnesota	Rochester	84,104
	Missouri	Columbia	80,911
		St. Joseph	86,915
		Springfield	152,929
	Nebraska	Lincoln	167,972
	North Dakota[a]		
	South Dakota	Sioux Falls	95,209
Southern Quadrant			
South Atlantic	Delaware[a]		
	Florida	Fort Lauderdale-Hollywood	620,100
		Gainsville	104,764
		Jacksonville	528,865
		Miami	1,267,792
		Orlando	428,003
		Pensacola	243,075
		Tallahassee	103,047
		Tampa-St. Petersburg	1,012,594
		West Palm Beach	348,753
	Georgia	Albany	89,639
		Atlanta	1,390,164
		Macon	206,342
		Savannah	187,767

Table A-1 (cont.)

Region	State	SMSA	Population
South Atlantic (cont.)	Maryland[a]		
	North Carolina	Asheville	145,056
		Charlotte	409,370
		Durham	190,388
		Fayetteville	212,042
		Greensboro-Winston-Salem	603,895
		Raleigh	228,453
		Wilmington	107,219
	South Carolina	Charleston	303,849
		Columbia	322,880
		Greenville	299,502
			95,209
	Virginia	Lynchburg	123,474
		Newport News-Hampton Bays	292,159
		Norfolk-Portsmouth	680,600
		Petersburg-Colonial Hts.	128,809
		Richmond	518,319
		Roanoke	131,436
	Washington, D.C.[a]		
	West Virginia	Charleston	229,515
East South Central	Alabama	Birmingham	739,274
		Gadsden	94,144
		Huntsville	228,239
		Mobile	376,690
		Montgomery	201,325
		Tuscaloosa	116,029
	Kentucky	Lexington	174,323
		Owensboro	79,486
	Mississippi	Biloxi-Gulfport	134,582
		Jackson	258,906
	Tennessee	Knoxville	400,337
		Nashville-Davidson	541,108
West South Central	Arkansas	Little Rock-N. Little Rock	287,189
		Pine Bluff	85,329
	Louisiana	Baton Rouge	285,167
		Lafayette	109,716
		Lake Charles	145,415

Table A-1 (cont.)

Region	State	SMSA	Population
West South Central (cont.)		Monroe	115,387
		New Orleans	1,045,809
		Shreveport	294,703
	Oklahoma	Lawton	108,144
		Oklahoma City	640,889
		Tulsa	476,945
	Texas	Abilene	113,959
		Amarillo	144,396
		Austin	295,516
		Beaumont-Port Arthur	315,943
		Brownsville-Harlingen	140,368
		Bryan-College Station	57,978
		Corpus Christi	284,832
		El Paso	359,291
		Fort Worth	762,086
		Galveston	169,812
		Laredo	72,859
		Lubbock	179,295
		McAllen-Pharr-Edinburg	181,535
		Midland	65,433
		Odessa	91,805
		San Angelo	71,047
		San Antonio	864,014
		Sherman-Denison	83,225
		Tyler	97,096
		Waco	147,553
		Wichita Falls	127,621
Western Quadrant			
Mountain	Arizona	Phoenix	967,522
		Tucson	351,667
	Colorado	Colorado Springs	235,972
		Denver	1,227,531
		Pueblo	118,238
	Idaho	Boise City	112,230
	Montana	Billings	87,367
		Great Falls	81,804
	Nevada	Las Vegas	273,288
		Reno	121,068

Table A-1 (cont.)

Region	State	SMSA	Population
Mountain (cont.)	New Mexico	Albuquerque	315,774
	Utah	Ogden	126,278
		Provo-Orem	137,776
		Salt Lake City	557,635
	Wyoming[a]		
Pacific	Alaska[a]		
	California	Anaheim-Santa Ana-Garden Grove	1,420,386
		Bakersfield	329,162
		Fresno	413,053
		Modesto	194,506
		Oxnard-Ventura	376,430
		Sacramento	800,592
		Salinas-Monterey	250,071
		San Bernadino-Riverside	1,140,166
		San Diego	1,357,782
		San Jose	1,064,714
		Santa Barbara	264,324
		Santa Rosa	204,885
		Stockton	290,208
		Vallejo-Napa	249,081
	Hawaii	Honolulu	629,176
	Oregon	Eugene	213,358
		Salem	136,658
	Washington	Seattle-Everett	1,421,869
		Spokane	287,487
		Tacoma	411,027

[a]These states have no SMSAs that were eligible for this study.

township for which the township government has solid waste authority has at least 2,500 population (by the 1970 Census).

Only 20 states have towns or townships. Three of these states (Rhode Island, Vermont, and North Dakota) have no SMSAs within the study universe (of 200 SMSAs). Two more states, Missouri and Nebraska, have no organized townships within their SMSAs, although they do have organized townships elsewhere within their boundaries. Of the remaining fifteen states with organized townships, five do not authorize their townships to provide for solid waste collection (Illinois, Indiana, Kansas, Minnesota, and South Dakota). Only ten states in the solid waste study universe have organized township governments with the authority to collect or arrange for the collection of solid waste. These are: Connecticut, Maine, Massachusetts, New Hampshire, New York, Wisconsin, Michigan, New Jersey, Ohio, and Pennsylvania. In the first six, the township form of government is called a "town." It may have one or more incorporated villages within its borders. In the last four states, the township form of government is called a "township." It may have one or more incorporated boroughs or villages within its borders (except in New Jersey). Connecticut is unique in that it has several consolidated borough-towns and city-towns; these are considered municipalities rather than townships by the census, and therefore the same convention was adopted for the purpose of this study. Two thousand sixty local jurisdictions were found in the 200 SMSAs that satisfied these two criteria.

Sample Bias Inherent in the Study Universe (200 SMSAs)

As is shown in Table A-2, the 200 eligible SMSAs in the study universe include approximately 14 percent of all local jurisdictions in the United States. Approximately 20 percent are in the 43 SMSAs which were excluded; the remainder are in rural areas. The obvious bias in the universe is against rural municipalities and townships, and very large cities and their suburbs, none of which is represented. The largest city in the study universe is Milwaukee, which had a population of 717,000 in 1970.

The UTS universe is further restricted: all local jurisdictions under 2,500 people are excluded. The remaining 2,060 local jurisdictions constitute approximately 30 percent of all organized noncounty local jurisdictions in the 200 SMSAs, and approximately 6 percent of all organized noncounty local jurisdictions in the United States.

Table A-2
Noncounty Local Jurisdictions in U.S.
(All population classes)

	200 SMSAs	243 SMSAs	Total U.S.
Central Cities	249	308	308
Other Municipalities	2,600	5,159	18,209
Township Govts.	1,968[a]	3,248	16,911
Total	4,817	8,715	35,428

[a]This number includes townships in 15 states in the 200 SMSAs. However, of this number 1,448 townships are authorized to provide for solid waste collection and thus are eligible for the study. 520 townships have no such authority and were excluded from the study.

Once the universe of 2,060 cities eligible for the study of waste collection practices was established, it was decided to include only some of these cities in the Universal Telephone Survey. This policy was intended to reduce the number of interviews and thus to allow for higher quality but more time-consuming interviewing techniques.

Cluster sampling was used for this purpose. As is shown in Table A-3, only 249 central cities are situated in the 200 eligible SMSAs. Since failure to call these cities would make identification of large cities with various characteristics for further analysis difficult, all 249 central cities were telephoned.

After these calls to central cities were made, 41 SMSAs were chosen for inclusion in the subsequent economic survey of cities with municipal collection. (This survey is described in Appendix B.) Before this latter survey could be conducted, suburban (noncentral) communities with municipal collection of

Table A-3
Noncounty Local Jurisdictions

	Solid Waste Collection Study Sample (jurisdictions ≥2500)	Solid Waste Collection Study Universe (jurisdictions ≥2500)	Total Number of Jurisdictions in 200 SMSAs (no size restriction)
200 SMSAs			
Central Cities	249	249	249
Satellite Cities and Towns	1,128	1,811	4,568[a]
Total	1,377	2,060	4,817
41 SMSAs			
Central Cities	51	51	51
Satellite Cities and Towns	445[d]	445	1,016[b]
Subtotal	496	496	1,067
159 SMSAs			
Central Cities	198	198	198
Satellite Cities and Towns	683	1,366	3,552[c]
Subtotal	881	1,564	3,750

[a]Number composed of 2,600 satellite cities, 1,448 townships having responsibility for solid waste collection, and 520 townships not having this responsibility.

[b]Number composed of 585 satellite cities, 339 townships having responsibility for solid waste collection, and 92 townships not having this responsibility.

[c]Number composed of 2,015 satellite cities, 1,109 townships having responsibility for solid waste collection, and 428 townships not having this responsibility.

[d]Does not include the one city which did not respond.

residential solid waste had to be identified. Therefore, as is shown in Table A-3, all 446 satellite cities and townships in these 41 SMSAs were included in the Universal Telephone Survey.

Every other (683 of 1366) satellite city and township in the remaining 159 SMSAs was called. Thus a total of 249 central cities and 1,129 satellite cities, or 1,378 cities, received the Universal Telephone Survey.

From a more detailed perspective, the Universal Telephone Survey sample represents all central cities in the 200 SMSAs, and approximately 80 percent of all central cities in the United States; approximately 30 percent of all satellite municipalities in the 200 SMSAs, and 4 percent of all satellite municipalities in the U.S.; 20 percent of all townships in the 200 SMSAs; and 2 percent of all townships in the U.S.

Data Collection

The Universal Telephone Survey provided basic descriptive data concerning solid waste collection practices in all local jurisdictions selected for the solid waste collection study sample (1,377). A telephone interview rather than a mail questionnaire was used, in order to obtain rapid results, a high response rate, a high accuracy rate, and first-hand impressions as to the reliability of the data. Approximately 60 percent of the local jurisdictions were telephoned by personnel from Public Technology, Inc. (PTI); the remainder were called by staff of the International City Management Association (ICMA). Professional staff from both these agencies participated in the design of the questionnaire, and then underwent training sessions in which mock interviews were conducted, followed by several monitored calls to noneligible cities. After several hundred calls to eligible cities had been made, a consistency check was performed to verify that the coding system was standardized among interviewers.

After all calls were completed, the results were edited in a multistage process to insure that the data given were accurate, that they were recorded correctly and consistently, and that they were transferred correctly from the interview forms to the computer file.

Description of Universal Telephone Survey

The Universal Telephone Survey was designed to acquire several different categories of information about the service arrangements for solid waste collection jurisdictions. The categories of data desired were the type of service arranger and service provider, the financing methods, the frequency and location of collection, and the existence of any limits on quantity collected per customer. These types of information were sought with respect to waste collected from single-family residential, multifamily residential, commercial, industrial, and institutional service recipients. Additional questions were directed toward separated collection of special types of residential waste, recent changes in organizational arrangements for collection of residential or commercial solid wastes, and the existence and extent of competition and choices in residential collection.

The Survey Instrument

The interview form was designed to allow for multiple responses wherever needed. If more than one service arrangement (municipal, contract) was used in a jurisdiction, one column of boxes was available for each type of collection. Similarly, multiple rows were available to record financing systems involving more than one type of revenue source. Within this framework, the interview form is fairly simple and straightforward, with the pages for responses folding out so that they are opposite the page with questions.

Instructions for Telephone Interview

See first page of interview form at end of this appendix.

Logistics of Data Collection

The total time for conducting the telephone survey was four months, commencing December 1974 and terminating March 1975. Approximately 300 calls were made on regular telephone lines during the month of December and the first week of January. During the period January 6-February 6, approximately 1,050 calls were made from WATS lines installed for this purpose at ICMA and PTI. Approximately 20 additional calls were made on regular lines during the period February 7-April 17 in order to complete the file. In all, ten interviewers participated in the Universal Telephone Survey—four at ICMA, six at PTI. All municipalities in the sample were successfully contacted and interviewed, with one exception—a jurisdiction in which no one ever answered the phone. In some cases, this required several attempts per jurisdiction, because many small local governments staff their offices only part time. Some townships do not maintain offices, thus necessitating a call to the home of the clerk or one of the supervisors or trustees of the township. In a few cases, the quality of the response was deemed inadequate, thus requiring a second call to another employee or official of the same jurisdiction.

It is important to note here that the interviewer insisted on speaking to an employee or official of the governing body of the local jurisdiction. This required some tenacity in cases where the government of the jurisdiction had no involvement in solid waste collection or disposal. One tactic frequently used in this situation was to ask to speak with someone in the government who lived in the jurisdiction and could speak, as a service recipient, with some knowledge of the service offered in the jurisdiction.

The important point here is that although the units of analysis in the Universal Telephone Survey were local *jurisdictions* (as for all of the surveys in the solid waste study), the respondents were local *governments*.

Number of Calls Attempted and Completed

In all, 1,377 successful calls were made, from a total sample of 1,378. This is a 99.9 percent response rate. ICMA's experience has been that mail surveys on this subject, of equivalent length and similar sample profile, rarely exceed a 50 percent response rate. Some improvement in response rate can be achieved in a mail survey by severely shortening the questionnaire (one page front and back), or by mailing to special samples of municipalities (for example, cities with the council-manager form of government).

Quality of Responses

The interviewers for the Universal Telephone Survey called each local government until a satisfactory interview was obtained (a "3" or better on a response quality rating scale of "1" to "5"). In a small number of cases, the interviewer, having exhausted all possibilities within the local government, called the private contractor or association of private contractors to whom the local government regularly referred solid waste collection calls.

In a substantial number of jurisdictions where collection was arranged directly between a private firm and the householders, the quality of responses was only mediocre—due to the lack of extensive knowledge on the part of the responding city official. In the majority of cases, however, particularly in those cases where the local government had some involvement in solid waste collection or regulation, the general quality of the response was good.

With very rare exception, the respondents were cooperative and patient in their responses. Many of the respondents exhibited interest in the survey and asked for a copy of the report. The telephone interview format proved extremely successful in terms of respondent attitude and as a means of finding knowledgeable respondents.

In general, respondents seemed quite knowledgeable about who collected refuse and who arranged for this collection for all types of service recipients (residents, multiple dwellings, commercial establishments). In cases where the municipality collected refuse, respondents were particularly knowledgeable about the service provided.

With respect to financing, if the jurisdiction assesses a tax or charges a fee for solid waste collection, this special tax or fee may or may not cover the full cost of the service. Yet, the respondent was rarely able and willing to say whether or not the service was also partially subsidized from general funds. (Chapter 6 discusses this issue.) Thus, in many cases, there is only a single response where there should be a multiple response. Adjustments in the analysis procedures eliminated any potential problem in this area.

Percentages reported for the fraction of households using each service arrangements are rough approximations at best, and are often untrustworthy, owing to lack of knowledge on the part of most of the respondents.

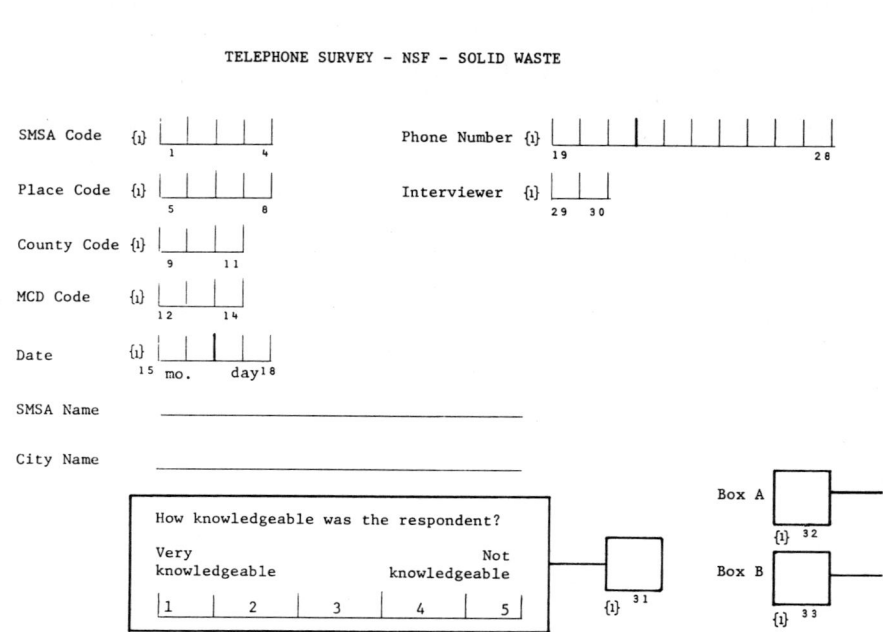

TELEPHONE SURVEY - NSF - SOLID WASTE

INSTRUCTIONS TO INTERVIEWERS FOR TELEPHONE SURVEY

1. Introduce yourself and ask to speak to someone knowledgeable about how garbage is collected in that city, town or village.

2. Instructions for Question I.

   a. Name each type of garbage (mixed household).

   b. Explain, using concrete examples, what we mean by that term.

   c. Suggest alternatives 1 through 9 to the respondent.

   d. Be sure to probe for multiple collections (a private contractor working for individuals and for the city should be entered twice as a number 2) and multiple alternatives (1, 2, etc.). Ask, "is everyone serviced by (fill in the alternative suggested by the respondent)?"

   e. For mixed household garbage, if the interviewee says that trash and wet garbage are collected separately, enter dry trash under mixed garbage and use wet garbage column.

3. For YES/NO questions, use 1 = YES, 2 = NO, 3 = DON'T KNOW.

I would like to ask you some questions about how garbage is collected in your city (village, town, etc.). I'd like to start by asking you about household collection.

I. Who collects mixed household garbage (define)? [Repeat for each refuse type.]

Codes: (Read)
1. Your municipality
2. A private hauling firm
3. The householder himself/establishment itself [If 3 answered for Question I, skip to Question III-E.]
4. The county
5. Another municipality
6. A special authority or special sanitation authority or district
7. A voluntary association (hslds., estabmnts.)
8. Other
9. Don't know

Are multiple dwellings treated the same as single households? [1 = YES; 2 = NO]...

If "NO," how many units must be present in a building before it is treated differently?...

II. Who arranges for services by (hires) this collector?

(Use same codes as in Question I above)
(If private firms are the collectors, ask if they are franchised, by whom, and enter the appropriate code.)

Contract/franchise

contract = 0
franchise = 1

III. For each garbage collector: [For each column entry in Question I.]

A. How is the service paid for? [If more than one, enter vertically for each column.]

1. General fund
2. A special property tax levy
3. Other special tax
4. A flat fee user charge collected by service decision maker
5. A flat fee user charge collected by garbage collector
6. A variable fee user charge collected by service decision maker
7. A variable fee user charge collected by garbage collector
0. Unknown/don't know

B. How often is the garbage picked up?

1. Once per week
2. Twice per week
3. More than twice per week
4. Once per month
5. Scheduled, but less frequently than once per month
6. On request
7. Various seasonal arrangements (only applied to mixed or wet garbage)
8. Various other arrangements

C. Where is the garbage picked up?

1. Curbside
2. Front yard
3. Backyard
4. Alley
5. Various pick-up sites
6. Don't know

D. Is there any limit on the amount of garbage picked up?

0. No limits
1. Limits on amount to be collected (quantity)
2. Excess collected with charge
3. Unknown/don't know

E. What percentage of households/multiple dwellings/commercial are serviced this way?

(Use three digits: e.g., 052 = 52%)

	RESIDENTIAL							
	Mixed household	SEPARATED			Other cd.	Other cd.	Multiple dwellings	
		Wet garbage	Leaves	Bulk				

Other codes:
1. Yard trash
2. Newspapers
3. Glass
4. Cans
5. Other

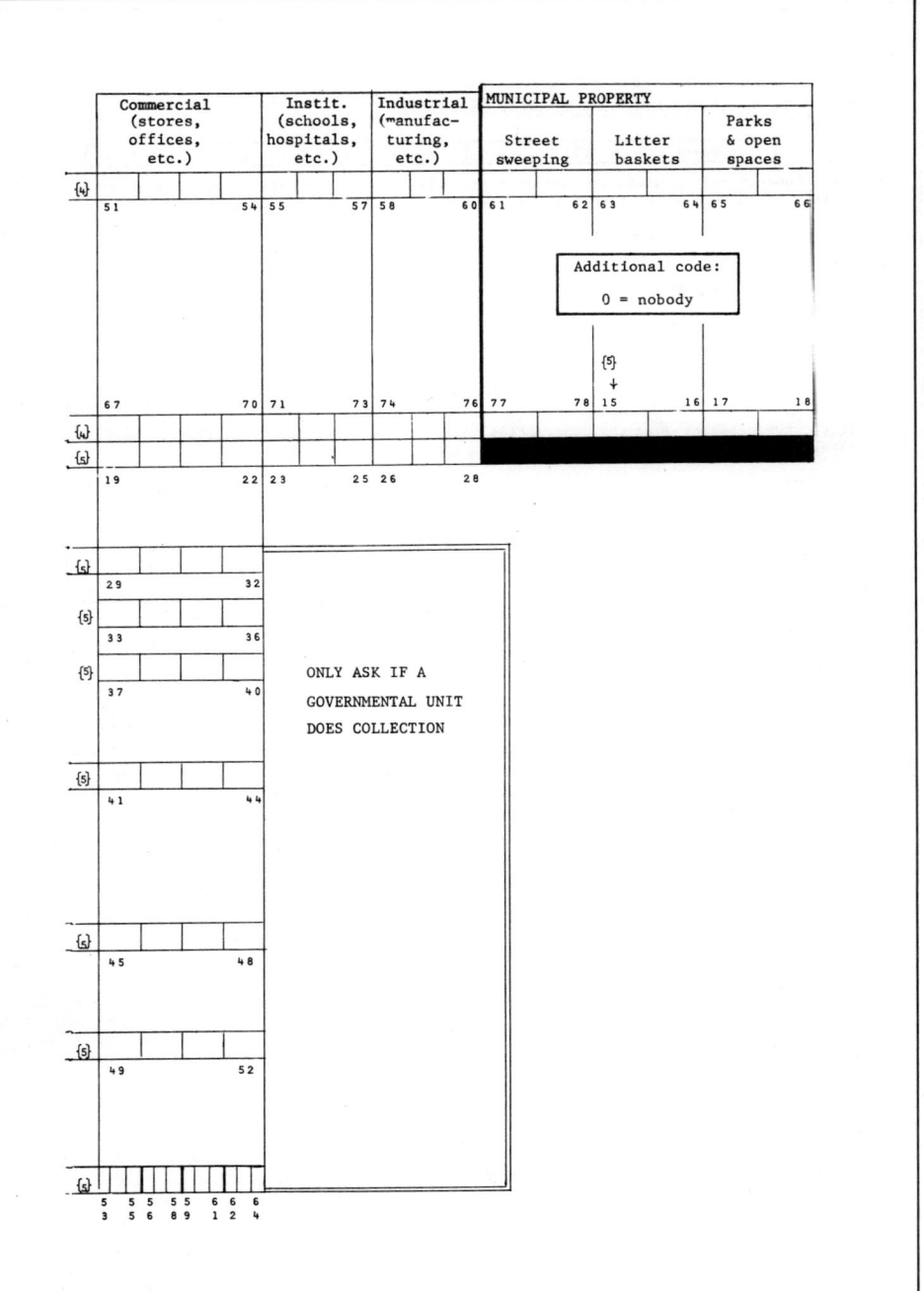

ADDITIONAL QUESTIONS AND ITEMS TO PROBE ON:

For Questions IV, VI, VII, and VIII use 1 = YES; 2 = NO; 3 = DON'T KNOW

IV. A few moments ago we discussed how both residential and commercial garbage were collected. Is this the same way it was collected 5 years ago or have there been changes?

    A. There have been residential changes..   65

    B. There have been commercial changes...   66

Describe changes completely (Include types and dates of changes).

_____

_____

_____

V. If the municipality does any <u>contracting to</u>, or has a <u>franchise with</u>, private haulers, please ask for:

    A. A copy of the contract or agreement to be mailed to: Solid Waste Project, Columbia University, New York, New York 10027.

        If we must write for agreement, get name and address of person to contact Check Box A on front.

    B. Check Box B if names and addresses filled in. Get names, addresses, and telephone numbers of contractors.

VI. Practically speaking, do any households in your municipality have a choice as to who will collect their garbage?..   67

    If "YES," about what percent of your households have this choice?..........   68

        1 = 10% or less   3 = 50% - 90%   5 = 100% (everyone)
        2 = 10% - 50%   4 = more than 90%   6 = don't know

VII. Do any households have a choice (for a fee) as to the following levels of service?

    A. Frequency of service...   69

    B. Location of pick-up...   70

    C. Any extra services (bulk, more cans, etc.)............................   71

VIII. Are all households required by local ordinance (law) to have garbage collection services?..   72

Name, title, and address of person you spoke to: _____

_____

Appendix B
Determining the Cost of Municipal Collection

Since city agencies were known to have unreliable information on the cost of municipal collection it was decided that cost data would be collected in person by interviewers. Hence, an economic survey form was designed for use by the interviewers. In addition, each interviewer filled out questionnaires on local solid waste regulation, local disposal practices, and on the effect of state and federal laws on waste management. This appendix describes the selection of cities visited and the methods used to complete the economic survey, and thereby to determine the cost of municipal collection.

Sample Selection

Selection of SMSAs for the Economic Survey

To reduce data collection expenses, it was decided to obtain economic data from three localities providing municipal solid waste collection services in each SMSA visited. Initial site visits tended to confirm the hypothesis that satellite cities often had the same organizational arrangement for solid waste collection as the central city. If the central city collects residential refuse municipally, it is likely that other cities in that SMSA will also have municipal collection forces. Thus, SMSAs were selected according to the organizational arrangement for solid waste collection that prevails in the central city.

Information about the organizational arrangement for solid waste collection in the central city of each of the 200 eligible SMSAs was obtained from two sources: responses to the ICMA/EPA solid waste survey—120 central cities—and telephone calls to the director of public works, the city manager, or the sanitation departments—80 central cities. To ensure that a sufficient number of satellite cities in each selected SMSA would have municipal collection, the universe of 200 central cities was stratified into those with 65 percent or more municipal collection (146) and those with other organizational systems. Central cities providing municipal collection services to fewer than 65 percent of the locality were excluded from the remainder of the selection procedure.

Central cities providing 65 percent or more of their municipality with municipal collection services were stratified by geographic region. The number to be selected from each region was calculated as the number of eligible central cities in that region times 40 (the desired number of SMSAs to be visited) over 146 (the total number of eligible central cities). Numbers were rounded off as necessary, to a total of 41 SMSAs visited. This is shown in Table B-1. Next,

Table B-1
Site Selection for SMSA Visits

	Stratification by Organizational Arrangement and Region				
Region	States	Total Eligible SMSAs	SMSAs with Municipal Collection[a]	Desired No. of SMSAs	Selected No. of SMSAs
West North Central	N.Dak., S.Dak., Minn., Neb., Iowa, Kan., Mo.	11	5	1.5	2
Middle Atlantic	Penn., N.J., N.Y.	19	16	4.8	5
East South Central	Ky., Tenn., Miss., Ala.	11	9	2.7	3
West South Central	Tex., La., Okla., Ark.	32	30	9.0	7
New England	Mass., Conn., R.I., Me., Vt., N.H.	21	15	4.5	4
East North Central	Ill., Ind., Ohio, Wis., Mich.	40	22	6.6	6
South Atlantic	Va., N.C., S.C., Ga., Fla., Md., Del., W.Va.	32	32	9.6	8
Mountain	Mont., Idaho, Wy., Nev., Utah, Col., Ariz., N.Mex.	14	9	2.7	3
Pacific	Alaska, Wash., Ore., Calif., Hawaii	20	8	2.4	3
Total		200	146	43.8	41

[a] In center city of SMSA.

SMSAs were selected at random from each region. The final sample is shown in Table B-2. These 41 selections constituted the sample of SMSAs visited for case studies.

Selection of Cities to be Visited

Once the 41 SMSAs were selected, individual cities to be visited had to be chosen. The Universal Telephone Survey was used as the instrument to make the selections. Every local jurisdiction ($\geq 2,500$ in population) within the 41 SMSAs was telephoned. After the telephoning for an entire SMSA was completed, the

Table B-2
SMSAs Selected for Site Visits, by Region

Region	State	SMSA	Municipalities Visited	Population
New England	Connecticut	New Haven	New Haven	137,707
			North Haven	22,194
			West Haven	52,851
		Waterbury	Waterbury	108,033
			Beacon Falls Town	3,546
			Naugatuck	23,034
	Maine	Portland	Portland	65,116
			South Portland	23,267
			Westbrook	14,444
	New Hampshire	Nashua	Nashua	55,820
Middle Atlantic	New Jersey	Jersey City	Bayonne	72,743
			N. Bergen Township	10,577
			Kearny	37,585
	New York	Rochester	Rochester	296,233
			Albion	5,122
			Medina	6,415
			Spencerport	2,929
	Pennsylvania	Johnstown	Johnstown	42,476
			E. Conemaugh	2,710
			Westmount	6,673
		Reading	Reading	87,643
			Kutztown	6,017
		Wilkes-Barre	Wilkes-Barre	58,876
			Edwardsville	5,633
			Newport Township	6,063
East North Central	Indiana	Muncie	Muncie	69,080
		South Bend	South Bend	125,580
			Plymouth	7,661
	Michigan	Flint	Flint	193,317
	Ohio	Springfield	Springfield	81,926
		Youngstown/Warren	Youngstown	139,788
			Warren	63,494
	Wisconsin	Milwaukee	Milwaukee	717,099
			Port Washington	8,752
			Wauwatosa	58,676

Table B-2 (cont.)

Region	State	SMSA	Municipalities Visited	Population
West North Central	Kansas	Wichita	Wichita	276,554
			Augusta	5,977
			Eldorado	12,308
	Missouri	Columbia	Columbia	72,691
			Centralia	3,618
South Atlantic	Florida	Miami	Miami	334,859
			Biscayne Park	2,717
			Miami Beach	87,072
			North Bay	4,831
		Pensacola	Pensacola	59,507
			Milton	5,360
		Tallahassee	Tallahassee	71,897
		West Palm Beach	West Palm Beach	57,375
			Lake Park	6,993
			Pahokee	5,663
	Georgia	Atlanta	Atlanta	496,973
			Buford	4,545
			College Park	18,203
			Lawrenceville	5,115
		Savannah	Savannah	118,349
			Garden City	5,741
			Port Wentworth	3,905
	North Carolina	Raleigh	Raleigh	121,577
			Fuquay-Varina	3,576
			Wake Forest	3,148
	Virginia	Norfolk/ Portsmouth	Norfolk	307,951
			Portsmouth	110,963
			Chesapeake	89,580
East South Central	Kentucky	Lexington	Lexington	108,137
	Mississippi	Biloxi/ Gulfport	Biloxi	48,486
			Gulfport	40,791
			Long Beach	6,170
	Tennessee	Nashville/ Davidson	Nashville/Davidson	448,003
			Gallatin	13,093
			Lebanon	12,492

Table B-2 (cont.)

Region	State	SMSA	Municipalities Visited	Population
West South Central	Arkansas	Little Rock	Little Rock	132,483
			Jacksonville	19,832
			Sherwood	2,754
	Oklahoma	Lawton	Lawton	74,470
	Texas	Abilene	Abilene	89,653
			Anson	2,615
			Hamlin	3,322
		Beaumont	Beaumont	115,919
			Pear Ridge	3,697
			Port Neches	10,894
		Brownsville	Brownsville	140,368
			Harlingen	33,503
			LaFeria	2,642
		Bryan-College Station	Bryan	33,719
			College Station	17,676
		Waco	Waco	95,326
			Robinson	3,807
			Woodway	4,819
Mountain	Arizona	Phoenix	Phoenix	581,562
			Mesa	62,853
			Scottsdale	62,907
	Montana	Great Falls	Great Falls	60,091
	New Mexico	Albuquerque	Albuquerque	243,751
Pacific	California	Fresno	Fresno	165,972
			Clovis	13,856
			Sanger	10,088
		Sacramento	Sacramento	254,413
			Lincoln	3,176
			Roseville	17,895
		San Diego	San Diego	696,769
			Imperial Beach	20,244
			Oceanside	40,494

names of all localities having municipal residential collection systems for 65 percent or more of their residents were written down on separate pieces of paper. Depending on the total number of cities with municipal collection in each SMSA, two or three cities were chosen at random from among the pile of papers. Thus, the central city and two or three satellite cities were chosen. Obviously, where only one (no) satellite city had municipal collection, the total number of municipalities to be visited within that SMSA would be two (one). In total, the 102 municipalities listed on Table B-2 were chosen. Tables B-3 and B-4 show the stratification of the sample by region and population class in comparison to the total universe of cities with municipal collection in the overall sample.

Data Collection

The Economic Survey

The economic survey instrument was a very detailed form in which each cost component of residential solid waste collection was listed. Figures were obtained from documentation of expenditures supplied by the city from budgets, general ledgers, controller's reports, bills, department expense statements, and so forth.

The first part yielded an overview of the service provided by the city. The second part was composed of five detail sheets on which municipal expenditures for all relevant categories were entered and totaled. The total was used to obtain efficiency measures such as cost per ton, cost per household, and so on. The actual form and instructions for computing the various cost components are reproduced at the end of this appendix.

Logistics of Data Collection

Visits were scheduled from January 6, 1975, through April 14, 1975. Depending on their size and location, one or two SMSAs were visited per week by a single person or by two persons.

Rate and Quality of Response

Data collection was frequently an arduous task since municipalities did not necessarily keep information in the format that was required for the survey form. Response rate and quality did vary from city to city.

Overall, the most difficult data to obtain were amount of waste collected, certain fringe benefit costs, and accurate maintenance costs. Further, it proved difficult in certain cases to separate out the portion of costs attributable to

Table B-3
Sample Universe of Cities with Municipal Collection, by Region and Population

Population	Region									Total	Percentage
	NE	MA	ENC	WNC	SA	ESC	WSC	MTN	PA		
2,500-9,999	7	47	27	2	54	18	40	11	18	224	40
10,000-49,999	18	25	37	3	30	8	25	7	11	164	29
50,000-99,999	12	7	14	3	17	3	16	7	3	82	15
100,000+	7	7	14	3	20	8	14	5	8	86	15
Total	44	86	92	11	121	37	95	30	40	556	100

Note:
NE = New England
MA = Mid-Atlantic
ENC = East North Central
WNC = West North Central
SA = South Atlantic
ESC = East South Central
WSC = West South Central
MTN = Mountain
PA = Pacific

Table B-4
Municipal Sample by Region and Population Class

Population	Region									Total	Percentage
	NE	MA	ENC	WNC	SA	ESC	WSC	MTN	PA		
2,500-9,999	1	8	2	2	11	1	8	1	1	35	34
10,000-49,999	3	3	0	1	1	4	4	0	5	21	21
50,000-99,999	4	2	4	1	5	0	4	3	0	23	23
100,000+	2	1	4	1	6	2	2	2	3	23	23
Total	10	14	10	5	23	7	18	6	9	102	100

Note:
NE = New England
MA = Mid-Atlantic
ENC = East North Central
WNC = West North Central
SA = South Atlantic
ESC = East South Central
WSC = West South Central
MTN = Mountain
PA = Pacific

residential solid waste collection. In both the very large municipalities and the very small ones, data collection was complicated because cost information was not available or was combined with other departmental costs. Thus, especially in large cities, municipalities with central vehicle operations did not keep accounts in terms easily transferable to our forms, vehicle insurance cost was not easy to secure where fleet insurance covered all municipal vehicles regardless of type, and the cost of billing customers for user charges were seldom kept separate from billing costs for water, sewer, and so forth.

In addition, certain special problems arose. (1) Some cities had changed their collection operations in the middle of the fiscal year. The officials were interested in what it was costing them now, and felt it was a waste of their time to look at the cost of a system no longer in effect. (2) Some cities had given large raises or expanded their operations since the end of the last fiscal year, thereby leading to the same attitude problem mentioned in (1). (3) Labor costs of very small cities were sometimes difficult to isolate, since the men worked only part time on refuse collection. In these cases, overtime, vacation, absentee rates, relief personnel, and so on, became more difficult to account for.

Because interviews were done in person, the ultimate response rate was nearly 100 percent on every question. Even if information could not be secured during the on-site visit, interviewers were responsible for tracking down all missing information, often with follow-up telephone calls. Forms were not turned in until fully completed.

Rigorous quality control measures were placed on the completed forms owing to their central position in the analysis. An individual who was not a part of the interview teams reviewed each form, making sure that there were no missing data and that the expenditures reported were reasonable. A number of consistency checks on the data were performed. If the data were inconsistent or missing, the individual who conducted the interview was notified and was responsible for calling back the locality to clear up or verify questionable data.

Upon completion of this review process all localities were sent letters of thanks and a copy of the completed cost analysis. The data on the forms were keypunched.

Instructions to Interviewer for Completing Municipal Collection of Residential Solid Waste Cost Survey

This document must be read thoroughly before beginning any data collection.

A. Familiarize yourself with the forms. They are divided into two main sections:

   1. Background Information and Summary Sheets
   2. Detail Sheets

B. The Background Information will be used in our econometric analysis (together with demographic and economic—such as level of income—data) to explain intercity variations in the cost of collection. Most of this information should be readily available. Request it and record it in Sections A through D of the Municipal Collection of Residential Solid Waste Cost Survey summary form.

   1. Check all responses to ascertain that they refer to the most recently completed fiscal year.

   2. Probe the interviewee to make sure that the number of trucks, employees, etc., refers *only* to those used for *residential* refuse collection. Alter the responses if the respondent has included trucks which operate separate routes for commercial or institutional customers. If the municipality collects, in the course of a regular residential route, from small commercial establishments (whose refuse conforms in quantity and type of containerization to that collected from households), then these accounts, and the trucks and crews they are serviced by, should be included in the Background Information Sheet. In short, exclude accounts which are dissimilar to residential accounts, whether in schedule of collection or quantity of refuse generated. Be sure to add any included commercial accounts to the number of households serviced.

   3. We do *not* want figures to reflect time and trucks used solely for collection of "unusual" types of residential solid waste—i.e., bulky items, separate collection for yard clippings or leaves, etc.

   4. Many of the responses must be estimated. Encourage the respondent to do so, where necessary.

5. Obtain a copy of the budget pages which refer to solid waste collection; they may be useful in clarifying the organization, structure, and costs.

C. The Detail Sheets require informed estimates to be made by the respondent. Make sure that these estimates are internally consistent.

D. The front portion of the Summary and Background Sheets should be filled in (with the exception of updated figures for population, etc.) prior to arrival at the interview. Updated numbers should be entered in the boxes provided only if their source is a notarized (approved by Bureau of the Census) or equivalent document. In cases where such information is available, enter the year of the census in the boxes directly beneath the entry.

E. Definitions which may be useful (numbers refer to Summary and Background Sheet items).

1. *Number of households serviced by the city* refers to households (living units) plus any commercial establishments which are serviced on regular residential routes, and which generate refuse at a rate similar in quantity and type of service required to that of households. Be sure not to collect accounts, rather than households. In the case of a city which collects from apartment houses, the number of multifamily accounts is probably equal to the number of apartment buildings rather than the number of households residing in those buildings (e.g., some of this can be obtained from census data—i.e., if the city collects from all single family units).

2. Number of different arrangements for residential refuse collection should equal the number of columns filled in for mixed household refuse in the Universal Telephone Survey.

3 & 4. Percentage of households serviced by city and percentage of households serviced at. Each of these sets of entries should add to 100 percent.

5. Enter the limit as the number of cans.

6. Percentages should add to 100 percent.

7 & 8. If the city has several disposal sites, at least one with scales and at least one without, the number 3 would be entered.

10. If the city provides only residential refuse collection services, and if disposal figures are reported separately in the budget, this number will agree with that of number 9.

12. Try to get the actual figures of revenues collected from households being included in the rest of the report. Number 12 should equal number 13 times number 1 times 12 months. If not, be suspicious and find out why not.

14c. The assessment ratio is usually widely known. Try to obtain a published source for this figure, however, or at least a knowledgeable response from the city assessor. Define.

17a. If the city is on a task completion incentive system, be sure the respondent understands you want hours of work for which the men are paid, which is usually greater than the hours they actually work.

17b. We want the nominal length of a regular work day, *not* the actual hours worked. In short, the number of hours a crew would have to work at its regular route before being entitled to overtime.

18. The number of crews working on an average work day should equal the number of truck shifts in an average work week (number 24b) divided by the number of days per week collection occurs (number 17c). It need not be a whole number.

    For example, suppose a city has two trucks operating once a week collection on two days of the week, each with one shift. There are then four truck shifts in an average work week. The number of crews working on an average work day is thus two.

    Households serviced per crew shift equals number 1 divided by the product of number 17c, number 18, and number 17b.

    Households serviced per paid crew hour equals:

    $$\frac{\text{Number of households serviced (number 1)}}{(\text{number 17b}) (\text{number 18}) (\text{number 17c})}$$

    Reasonable ranges for these figures for typical suburban curbside service:

    Households serviced per crew shift: 300-700
    Households serviced per crew hour: 30- 90

19. Average crew size may often not be a whole number. Calculate as:

    $$\frac{\text{Number of one-man crews plus number of two-man crews times two plus number of three-man crews times}}{\text{total number of crews operating on an average work day (number 18)}}$$

20. Probe to get an accurate number here. Where seniority leads to more vacation days, this will not be equal for all crew men and the average may not be an integral number of days.

21. We want the total absentee rate here; i.e., the percentage of collectors and drivers on the staff not present for work (for all reasons, *including vacation*) on an average day.

22. Describe the incentive system briefly in words. Do not fill in the boxes (which will be coded later).

25. This number should be less than or equal to the sum of the product column 2 and 4 of Detail Sheet number 5.

    Reasonable Ranges:

37. $18-$70

38. $15-$40

39. $3-$8

40. $300-$700 (See instructions for line 18 for calculation method.)

Instructions for Detail Sheet 1: Salaries for Residential Solid Waste Collection

A. To be completed during interview:

1. Fill in Column *1*, Column *3, and either* Column *5* or *6*.

2. Include all personnel below the level of mayor or city manager who are involved in residential solid waste collection (RSWC). Check to make sure salaries listed apply to the correct fiscal year.

3. Go over the list of employees to make sure no one was omitted. If a salary changed during the year, calculate the weighted average salary.

4. Entries in Column 3 should be in decimal percentage. For example, if the division manager spends 30 percent of his time supervising activities in residential refuse collection the entry in Column 3 would read: "0.30"

5. Entries in Column 1 need not be whole numbers. For example, if a truck driver earning an annual salary of $10,000 worked only 4 months of the year under consideration, he should be entered as 1/3 of an employee (4 months out of 12) in Column 1.

B. To be completed after interview:

1. Enter totals in Column 7 as follows:

    (a) Where an annual salary is reported, multiply

    (Column 1) (Column 3) (Column 5)

    and enter the result in Column 7.

    (b) Where an hourly rate is reported, calculate the annual salary from the hourly rate and the number of hours worked per year. Then proceed as in 1(a) and enter the result in Column 7.

    (c) Overtime may be reported as a lump-sum amount; multiply this amount by Column 3 and enter in Column 7.

    (d) Be sure all percentages are in decimal form for these calculations, e.g., 10 percent should be carried as 0.10 in your calculations.

2. Total the entries in Column 7. This figure should be recorded on the Summary on line 30.

3. Complete Column 4 as the product of Columns 1 and 3.

Instructions for Detail Sheet 2: Fringe Benefits

A. To be completed during interview:

1. All figures are to reflect city contributions only, as fringe benefits to the personnel listed on Detail Sheet 1.

2. Column 2, type of employees covered, is critical. For example, if the city provides uniforms to truck drivers and collectors only, at a cost of $100/man/year, this fact must be noted, or else the calculation of the total will be seriously biased. Use the space under comments and notes to make additional notations about employees covered.

3. Column 3 should contain either a decimal percentage (e.g., social security written as 0.0585) of total expenditures per year (e.g.,

$2,000/year/man), or the specific amount spent per man/year (e.g., $100 for uniforms, written as $100/year/man).

B. To be completed after interview:

1. Check the respondent's answers to make sure they are complete and refer to the proper fiscal year.

2. Calculate the total for social security as follows:

   Check to see if any personnel listed on Detail Sheet 1 earn more than $13,200 annually. If so, adjust the total salaries figure calculated by treating these individuals as though they earned $13,200 and repeating the calculation of total salaries of refuse collection operation personnel. Multiply the adjusted total times 0.585 and enter the product in Column 5. Be sure, however, to note what base salary was covered by social security during the fiscal year under consideration. It is currently $13,200. For the first half of 1974 it was $12,000; during the last half of 1973, it was $10,800.

3. Where fringe benefits, other than social security, are reported as a percentage of salary, multiply the decimal percentage times the original (unadjusted) total salary figure calculated on Detail Sheet 1 and enter the product in Column 4 on Detail Sheet 2.

4. Where fringe benefits other than social security are reported as a lump sum item, inquire whether these benefits apply to all personnel listed in Detail Sheet 1. If they apply only to one category of personnel (e.g., uniforms only to collectors and drivers), then multiply the total amount times the percentage entered on Detail Sheet 1 for these employees and enter the product in Column 4 on Detail Sheet 2.

   If benefits apply to all personnel, then do the following: From Detail Sheet 1, calculate what total salaries would be if all personnel spent 100 percent of their time on residential refuse collection. Call this X.

   Then take the ratio of total salaries of residential collection operation personnel to X, multiply the ratio times the total amount listed for the fringe benefit, and enter the product in Column 4 on Detail Sheet 2.

5. Where the fringe benefit is reported as dollars per man per year, again inquire whether the benefit is applicable to only one category of personnel. If so, multiply the number of such employees by the

fraction of their time spent on residential refuse collection (from Detail Sheet 1) then multiply the $ per man per year and enter the product in Column 4 on Detail Sheet 2.

If the benefit applies to all personnel, calculate the number of full-time equivalent personnel in Column 4 of Detail Sheet 1 by multiplying each entry in Column 1 by the corresponding entry in Column 3, and entering the product in Column 4. The sum of Column 4 is the number of full-time equivalent personnel. Enter this in the Summary on Line 15. Multiply this number by $ per man per year and enter the product in Column 4 on Detail Sheet 2.

6. Total the entries on Column 4 and enter the sum in your Summary Sheet under fringe benefits (Line 31).

7. Divide total fringe benefits by total salaries of residential collection personnel and enter the result at the bottom of Detail Sheet 2.

Instructions for Detail Sheet 3: Other Operating Expenses

A. To be completed during interview:

Make sure all data is specific to the most recently completed fiscal year.

Percentages should be entered in decimal percentage form (30 percent is entered as 0.30).

1. Vehicle Operation and Maintenance.

    Try to obtain these figures broken down by category. Be sure to make the respondent estimate the expenses for trucks used in residential refuse collection, i.e., if equipment expenses are $1,000,000, but vehicles are used only 50 percent of the time for residential refuse collection, then enter $1,000,000 in Column 2 and 0.50 in Column 3. In some cases, if trucks are used for other kinds of collection, it may be necessary to calculate the fraction of total truck hours which are used only for residential collection.

2. Division Office Expenses

    Try to obtain bills for these items. By division office expenses, we mean office expenses incurred to support the personnel listed on Detail Sheet 1. Include purchases of radios, office equipment, etc.

If the figures are not available by category, take nonlabor expenses of the division as the item to be entered in Column 2 (from the budget). In Column 3, enter grand total from Detail Sheet 1 divided by total salaries of the division (obtained from the budget). As you will not know what the grand total from Detail Sheet 1 is at this time, note total salaries of the division under comments and note for later use.

3. Billing

   If the city bills its residents, ask what the costs are, the applicable percentage attributable to residential solid waste collection billings, and record these. Then note (comments and notes) how these numbers are derived.

4. Insurance

   Try to see the city's insurance policy and to pick out from it the items applicable to the residential collection force.

5. Annual Interest

   If the city pays interest on equipment or buildings for use by the residential collection force, record this amount here.

6. Contractual Services

   Enter here the cost of any contractual services attributable to residential solid waste collection.

7. Miscellaneous—Other operating costs that are not included elsewhere in Detail Sheet 2 or Detail Sheet 3.

B. Portion to be completed after the interview.

   Except as noted for division office expenses, fill in Column 4 for each line item with the product of the entries in Columns 2 and 3.

   Sum the entries in Column 4 and enter on the bottom of the page and also on the Summary on Line 32.

   Enter the totals for each category on the Summary Sheet in Lines 32a through 32g.

Instructions for Detail Sheet 4: Overhead Costs

A. To be completed during interview:

1. Look through revised (completed) city budget with interviewee and identify those departments which provide general overhead support for residential solid waste collection. List them in Column 1.

2. Enter in Column 2 the total salaries allocated with each of the departments.

3. At the same time, enter in Column 3 the total nonsalary expenses (other than capital expenses) associated with each of these departments. *Exclude debt service.*

B. To be completed after the interview:

1. *Overhead Salaries for RSWC*

    a. Sum Column 2 ("Salaries and wages") on Detail Sheet 4, and enter in *Total* blank under Column 2. Make sure total overhead salaries and wages are exclusive of any fringe benefits (i.e., social security, etc.).

    b. Multiply *total* of Column 2 by the ratio calculated at bottom of sheet 1 (RSWC Fraction of Total City Salaries). Enter the product on Line B1 and on the Summary Sheet on Line 34a.

2. *Fringe Benefits*

    Multiply the answer in Line B1 by the fringe benefit fraction calculated at the bottom of Detail Sheet 2. Enter this figure on Line B2 and on the Summary Sheet on Line 34b.

3. *Other Overhead Costs for RSWC*

    a. Sum Column 3 ("Other than salaries and wages") and enter in *Total* blank under Column 3.

    b. Multiple *total* of Column 3 by the ratio calculated at bottom of Sheet 1 (the same ratio used in Step 1b). Enter the product on Line B3 and on the Summary Sheet on Line 34c.

Instructions for Detail Sheet 5: Equipment Cost

Make sure all data are specific to the most recently completed fiscal year.

A. To be completed during interview:

1. *Column 1* should contain a brief description of the type of vehicle, for example: 1968 20-cubic-yard read loader packer. Different years should be on different lines. Note that sedans, pickup trucks, etc., may be used by supervisory personnel. Do not forget to include all backup vehicles used to any extent for residential solid waste collection. Also be aware that sometimes a new chassis is purchased and an old compaction unit is mounted on it. Be sure to identify such cases properly.

2. *Column 2* should contain the number of such vehicles owned (or leased) by the city.

3. *Column 3* should contain, in dollars, the purchase price paid for the vehicle.

4. *Column 4* should contain the percentage of time the vehicle is used for residential refuse collection, written as a decimal percentage. For example, if the truck is used half the time for residential refuse collection, the entry would be *0.50*.

B. To be completed after interview:

1. *Column 5* should be calculated as the product of Columns 2, 3, and 4.

2. *Column 6:* Enter in Column 6 x 20 percent of each entry in Column 5 which represents equipment purchased in 1970 or more recently. Make no entries for equipment purchased in 1969 or earlier. Calculate the sum of the entries in Column 6 and enter it at the bottom and on the Summary on Line 34.

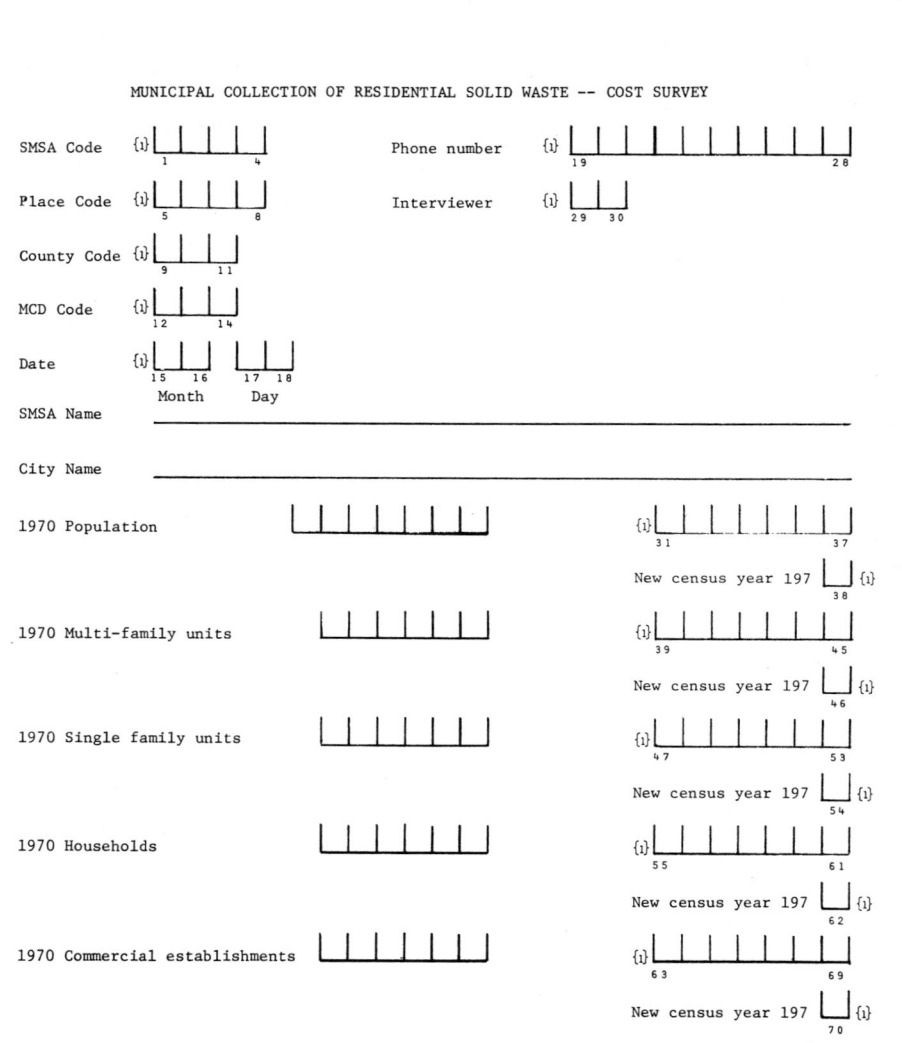

BACKGROUND INFORMATION

I. SERVICE DESCRIPTION

1. Number of households (and commercial establishments of a household nature) serviced by the city on a regular residential type route.................................... {1} ☐☐☐☐☐☐

   What is the accuracy of the response to Question 1? (Rank from 1 = most accurate to 5 = least accurate) {1} ☐

2. Number of different arrangements existing for household collection... {1} ☐

3. Percent of households serviced by city:

   a. Once a week... {2} ☐☐☐ ⎫

   b. Twice a week... {2} ☐☐☐ ⎬ %

   c. "Other" times per week (specify number of times)....... {2} ☐☐ ☐☐☐ ⎪

   d. Other (specify)_____ .... {2} ☐☐ ☐☐☐ ⎭

4. Percent of households serviced at:

   a. Curbside... {2} ☐☐☐

   b. Back yard.. {2} ☐☐☐

       1 = can returned   3 = can not returned
       2 = "barrel" system   4 = other _____ .. {2} ☐

   c. Alley.. {2} ☐☐☐

   d. Other (specify)_____ .... {2} ☐☐ ☐☐☐

5. Is there an upper limit on the number of cans that will be collected? (1 = YES, 2 = NO).................................... {2} ☐

   If "1," what is this limit?................................ {2} ☐☐

6. What percent of the refuse is collected in:

   a. Returnable containers (cans)...................... {2} ⎣__|__|__⎦
                                                              49   51

   b. Non-returnable containers (bags).................. {2} ⎣__|__|__⎦
                                                              52   54

7. Amount of residential solid waste collected annually by municipality:

   a. i. Tons.. {2} ⎣__|__|__|__|__|__|__⎦
                                                              '55   61

       ii. Weighed = 1; Estimated = 2; Both = 3.......... {2} ⎣__⎦
                                                              62

   B. i. Cubic yards................................... {2} ⎣__|__|__|__|__|__|__⎦
                                                              63   70

       ii. Measured = 1; Estimated = 2; Both = 3......... {2} ⎣__⎦
                                                              71

       iii. Compacted = 1; Not compacted = 2............. {2} ⎣__⎦
                                                              72

II. FINANCIAL INFORMATION

8. Start of the fiscal year for which data are obtained..... {2} ⎣__|__⎦ 7 ⎣__⎦
                                                              73 74   75
                                                              month   year

9. A. Municipal expenditures (as reported in city budget)
      for solid waste collection........................... {3} ⎣__|__|__|__|__|__|__|__⎦
                                                                 15   22

   B. Does this figure include the following? (1 = YES; 2 = NO)

      a. Disposal expenditures............................ {3} ⎣__⎦
                                                                23

      b. Capital outlays.................................. {3} ⎣__⎦
                                                                24

      c. Non-residential expenditures..................... {3} ⎣__⎦
                                                                25

10. What is the municipal expenditure for residential solid
    waste collection only (excluding capital outlays)?....... {3} ⎣__|__|__|__|__|__|__|__⎦
                                                                   26   33

11. What are total municipal expenditures for solid waste
    disposal, as reported in the city budget?.............. {3} 34   41

12. What are total municipal revenues for <u>residential</u> solid
    waste collection?...................................... {3} 42   49

13. If a fee is collected from households, what is the cost
    per household per month?............................... {3} 50   53

14. If a special property tax is collected:

    a. What is the total amount collected?............... {3} 54   61

    b. What is the rate in mills/$1,000 valuation?....... {3}   62 63

    c. What is the assessment ratio?..................... {3}   64 66

III. PERSONNEL INFORMATION

*15. What is the number of full-time equivalent employees in
    residential collection (from detail sheet #1)?......... {3} 67   71

16. What is the average monthly wage of a collector
    (excluding drivers)?................................... {3} 72   75

17. A. What are the number of paid hours per work week per
       collector (excluding drivers and overtime)?......... {3} 76 77

    B. How long is a crew shift in hours?.................. {4} 15 17

    C. How many days of the week does collection of residen-
       tial refuse occur (e.g., 4.5, 5.0, 6.0, etc.)........ {4} 18 19

18. How many collection crews work on an average work day?.. {4} 20 23

19. What is the average crew size (including driver)?...... {4} 24 25

20. What is the average number of vacation days per
    collector per year (including driver)?................. {4} 26 27

21. What is the annual average absentee rate for collectors
    and drivers?... {4} 28 30

22. Are collection crews on an incentive system?
    (1 = YES; 2 = NO)...................................... {4} 31

    If "YES," describe this system _____
    _____ {4} 32 33

23. What percentage of the residential collection crews have permanent civil service status (including drivers)?........................ {4} ⎵⎵⎵ 34 36

24. Are any members of the residential collection crews unionized? (1 = YES; 2 = NO).. {4} ⎵ 37

   If "YES," which of the following are present:.................... {4} ⎵ 38

      1 = AFSCME
      2 = Local teamsters
      3 = Other (specify) _____ {4} ⎵ 39

IV. EQUIPMENT INFORMATION

25. A. How many trucks collect residential refuse on an average day?.... {4} ⎵⎵⎵ 40 42

    B. How many truck shifts operate in an average week?................ {4} ⎵⎵⎵⎵ 43 46

26. What percentage of trucks used for residential collection are:

   a. i. Rear loader... {4} ⎵⎵⎵ 47 49

       ii. Front loader... {4} ⎵⎵⎵ 50 52

       iii. Side loader... {4} ⎵⎵⎵ 53 55

       iv. Other (specify)_____ ... {4} ⎵⎵⎵ 56 58

   b. Compactors.. {4} ⎵⎵⎵ 59 61

27. How many miles, on the average, does a residential collection truck drive per working day?... {4} ⎵⎵⎵ 62 64

28. What is the disposal cycle time (in minutes) for a residential collection truck (route to disposal point to route)?................ {4} ⎵⎵ 65 66

29. What percent of residential collection trucks dump at landfills?.... {4} ⎵⎵⎵ 67 69

*30. What is the average capacity of a residential collection truck (in cubic yards)?... {4} ⎵⎵⎵ 70 72

COMPLETE THIS PAGE FROM DETAIL SHEETS

*V. SUMMARY OF ANNUAL RESIDENTIAL COLLECTION COSTS

*31. Total salaries for residential collection personnel..... {$} |_|_|_|_|_|_|_|_|
  15   22

*32. Total fringe benefits................................. {$} |_|_|_|_|_|_|_|
  23   29

*33. Other operating expenses............................. {$} |_|_|_|_|_|_|_|_|
  30   37

    *a. Vehicle operation and maintenance............... {$} |_|_|_|_|_|_|_|_|
  38   45

    *b. Division office expenses........................ {$} |_|_|_|_|_|_|_|
  46   52

    *c. Billing expenses............................... {$} |_|_|_|_|_|_|_|
  53   59

    *d. Insurance expenses............................. {$} |_|_|_|_|_|_|_|
  60   66

    *e. Interest expenses.............................. {$} |_|_|_|_|_|_|_|
  67   73

    *f. Contractual services........................... {6} |_|_|_|_|_|_|_|
  15   21

    *g. Miscellaneous (other) operating expenses........ {6} |_|_|_|_|_|_|_|
  22   28

*34. Overhead costs....................................... {6} |_|_|_|_|_|_|_|
  29   35

    *a. Labor.. {6} |_|_|_|_|_|_|_|
  36   42

    *b. Fringe benefits................................ {6} |_|_|_|_|_|_|_|
  43   49

    *c. Other overhead costs........................... {6} |_|_|_|_|_|_|_|
  50   56

*35. Depreciation of vehicles............................. {6} |_|_|_|_|_|_|_|
  57   63

*36. Total costs (31 + 32 + 33 + 34 + 35).................. {6} |_|_|_|_|_|_|_|
  64   70

VI. PERFORMANCE MEASURES

37. Cost per year per household serviced (#36/#1)...................... {7} ⊔⊔⊔
    15   18

38. Cost per year per ton collected (#36/#7a.i.)....................... {7} ⊔⊔⊔
    19   22

39. Cost per year per cubic yard collected (#36/#7b.i.)................ {7} ⊔⊔⊔
    23   26

40. Number of households serviced per crew shift....................... {7} ⊔⊔⊔
    27   30

Detail Sheet #1: Salaries for Residential Solid Waste Collection (RSWC)

	Column 1	Column 2	Column 3*	Column 4	Column 5	Column 6**	Column 7
	Number of employees	Title of position	% of time spent on RSWC	Number of full-time equivalent personnel	Annual salary	Hourly rate (if applicable) and hours worked/yr.	Total
A.		Supervisory Personnel					
a.	_____	Commissioner........	_____	_____	_____	____/____	_____
b.	_____	Division manager....	_____	_____	_____	____/____	_____
c.	_____	Supervisor..........	_____	_____	_____	____/____	_____
d.	_____	Dispatcher..........	_____	_____	_____	____/____	_____
e.	_____	Foreman.............	_____	_____	_____	____/____	_____
		Other (specify)					
f.	_____	_____...	_____	_____	_____	____/____	_____
g.	_____	_____...	_____	_____	_____	____/____	_____
h.	_____	_____...	_____	_____	_____	____/____	_____
B.		Operations Personnel					
a.	_____	Truck dirver.......	_____	_____	_____	____/____	_____
b.	_____	Collector..........	_____	_____	_____	____/____	_____
		Other (specify)					
c.	_____	_____...	_____	_____	_____	____/____	_____
d.	_____	_____...	_____	_____	_____	____/____	_____
e.	_____	_____...	_____	_____	_____	____/____	_____
f.	_____	_____...	_____	_____	_____	____/____	_____
g.	_____	_____...	_____	_____	_____	____/____	_____
h.	_____	_____...	_____	_____	_____	____/____	_____
C.		Support Personnel					
a.	_____	Secretary..........	_____	_____	_____	____/____	_____
b.	_____	Clerk..............	_____	_____	_____	____/____	_____
c.	_____	Relief personnel....	_____	_____	_____	____/____	_____
d.	_____	Janitorial staff....	_____	_____	_____	____/____	_____
e.	_____	Temporary...........	_____	_____	_____	____/____	_____
f.	_____	Summer..............	_____	_____	_____	____/____	_____
		Other (specify)					
g.	_____	_____...	_____	_____	_____	____/____	_____
h.	_____	_____...	_____	_____	_____	____/____	_____
D.		Overtime............	_____	_____	_____	____/____	_____

Grand total RSWC..................................... _____

RSWC Salaries as Fraction of Total City Salaries.................. _____

COMMENTS AND NOTES: (use reverse side)

* Exclude any time spent on activities such as street cleaning, snow removal, leaf collection, commercial collection (if done on routes seperate from regular residential collection), bulk collection and disposal.

** If hourly rates given, be sure to determine how many hours are worked per year times hourly rate to obtain annual salary.

Detail Sheet #2: Fringe Benefits Attributable to Personnel Listed on Detail Sheet #1

Column 1	Column 2	Column 3	Column 4
Benefit	No. of employees covered	(%of salary, total amount, or $/man/year - specify which) Figure/Specification	Total
A. Social security............	_____	_____	_____
B. Insurance:			
a. Health and hospital.....	_____	_____	_____
b. Dental.................	_____	_____	_____
c. Life...................	_____	_____	_____
d. Disability.............	_____	_____	_____
e. Workmen's compensation..	_____	_____	_____
f. Unemployment compensation...................	_____	_____	_____
C. Retirement fund...........	_____	_____	_____
D. Uniforms and cleaning.....	_____	_____	_____
E. Safety equipment..........	_____	_____	_____
F. Longevity or bonus pay....	_____	_____	_____

    Grand Total _____

Fringe benefits as fraction of salary divided by grand total of DS#2 divided by grand total of DS#1................................... _____

COMMENTS OR NOTES:

Deatil Sheet #3: Other Operating Expenses

Column 1	Column 2	Column 3	Column 4
Category	Total annual expenses ($)	% applicable to residential solid waste collection	Total

I. Vehicle Operation and Maintenance
    A. Tires..........................
    B. Fuel...........................
    C. Oil............................
    D. Lubrication....................
    E. Maintenance labor..............
    F. Parts and materials............
    G. Other vehicle operating expenses (such as utilities for maintenance)..........................
    H. Total (if not available by category).......................

II. Division Office Expenses
    A. Telephone......................
    B. Utilities......................
    C. Office supplies................
    D. Rent...........................
    E. Communication equipment........
    F. Other _____

III. Billing............................

IV. Insurance
    A. Vehicles:
        1. Collision....................
        2. Liability....................
        3. Other _____
    B. Personnel:
        1. Liability....................
        2. Bonding.....................
    C. Property:
        1. Fire........................
        2. Theft.......................
        3. Other _____
    D. Damage claims paid.............

V. Interest...........................

VI. Contractual Services...............

VII. Miscellaneous/Other operating costs not included elsewhere............

                                                Grand Total _____

COMMENTS OR NOTES: (use reverse side)

Detail Sheet #4: Overhead Costs

A. Portion to be completed during interviews:

	Column 1	Column 2 Salaries and wages	Column 3 Other than salaries & wages
1.	Name of overhead agency		
	a. Mayor/Manager..................................	_____	_____
	b. Finance, Comptroller, Budget, Treasurer.........	_____	_____
	c. Clerk...	_____	_____
	d. Legal/Attorney..................................	_____	_____
	e. General Services................................	_____	_____
	f. Data Processing.................................	_____	_____
	g. Purchasing......................................	_____	_____
	h. Personnel.......................................	_____	_____
	i. _____ ...	_____	_____
	j. _____ ...	_____	_____
	k. _____ ...	_____	_____
	TOTAL	_____	_____
2.	Total non-salary expenditures for all city departments........		_____
3.	Total salary expenditures for all city departments..	_____	

B. Portion to be completed after interview:

1. Overhead salaries for RSWC........................ _____

2. Fringe benefits associated with the above.......... _____

3. Other overhead costs for RSWC..................... _____

Detail Sheet #5: Equipment Cost

Column 1	Column 2	Column 3	Column 4	Column 5	Column 6
Type of vehicle (Year, capacity, loading place, packer or not) or other large capital equipment	Number of vehicles	Acquisition cost ($) per vehicle	% of time used for residential refuse collection	(#2x#3x#4) total	20% of column 5 for vehicles purchased in the 1970's

Grand total _____

(Enter on Summary Page)

COMMENTS OR NOTES:

Appendix C
Determining the Prices
of Private Firms

Information about the price of refuse collection had to be collected for three types of firms: firms with municipal contracts; firms with franchises; and firms that sell their services to individual households.

Survey Strategy

For each type of situation, a different survey strategy was used. In all cases a target group of cities with each arrangement was selected using the results of the Universal Telephone Survey. Table C-1 shows the regions and populations of this chosen sample of 337 cities. These cities were picked to distribute the sample over the different regions and population groups as evenly as possible. As is shown in Table C-2, 125 of 272 contract cities identified by the Universal Telephone Survey were selected along with 87 of 99 franchise cities and 125 of 432 cities with private collection.

Cities With Municipal Contracts

All 125 contract cities selected for the cost analysis were sent a mail survey requesting information from government officials about the price the city paid their contractor(s) for refuse collection. This survey, the "Survey of Municipal Contract Procedures for Residential Solid Waste Collection" (MCON) is reproduced at the end of this appendix.

In all cases where the MCON was sent out, the contracting firm was telephoned and given the "Private Contractor Survey Questionnaire" (CON). This questionnaire is also reproduced at the end of this appendix.

Since cities were to be eliminated from the cost analysis if either the MCON or the CON was not completed, in order to increase the sample of contract cities the fourteen cities for which the CON but not the MCON were filled out were telephoned to fill out the MCON. Officials in all these cities were asked to send copies of their contracts and, in cases where these were received, they were used to confirm responses to the MCON and CON questionnaires.

Table C-2 shows that both the MCON and the CON were returned by 92 cities. (Note, however, that only 70 of these provided *all* the necessary cost data.) The regional and population breakdown of these cities is shown in Table C-3.

255

Table C-1
Distribution of Cities Initially Selected for Analysis

Population	Region									Total
	NE	MA	ENC	WNC	SA	ESC	WSC	MTN	PA	
Contract										
2,000-9,999	3	15	15	2	4	1	5	4	4	53
10,000-49,999	9	20	12	0	0	0	1	3	6	51
50,000-99,999	1	6	2	0	0	1	2	0	1	13
100,000+	0	2	2	0	0	0	0	0	4	8
Total	13	43	31	2	4	2	8	7	15	125
Franchise										
2,000-9,999	0	8	7	0	5	0	5	0	17	42
10,000-49,000	0	2	2	0	4	0	1	1	21	31
50,000-99,999	0	0	1	0	0	0	1	2	8	12
100,000+	0	0	0	0	0	0	0	1	1	2
Total	0	10	10	0	9	0	7	4	47	87
Private										
2,000-9,999	5	27	20	0	5	2	0	3	1	63
10,000-49,999	5	16	21	0	0	0	0	1	1	44
50,000-99,999	2	1	4	0	0	2	0	2	1	12
100,000+	2	0	1	0	0	2	0	1	0	6
Total	14	44	46	0	5	6	0	7	3	125

See Appendix A for region code and list of states in each region.

Table C-2
Selection of Cities with Private Service Providers, for the Economic Analysis

Arrangement	No. of Cities in UTS	No. of Cities Selected	No. of Cities Where Government Officials Responded	No. of Cities Where Contractors Responded	No. of Cities Where Householders Were Surveyed	No. of Cities Reporting	No. of Cities Reporting All Cost Data
Contract	272	125	99	95	NA	92	70
Franchise	99	87	NA	59	65	59	56
Private	432	125	NA	87	97	87	87
Total	803	337				238	213

Table C-3
Distribution of Cities Analyzed

Population	Region									Total
	NE	MA	ENC	WNC	SA	ESC	WSC	MTN	PA	
Contract										
2,000-9,999	3	10	8	0	3	2	3	3	2	34
10,000-49,999	8	16	10	0	0	0	1	2	3	40
50,000-99,999	1	4	3	0	0	0	2	0	1	11
100,000+	0	2	2	0	0	0	0	0	3	7
Total	12	32	23	0	3	2	6	5	9	92
Franchise										
2,000-9,999	0	4	4	0	1	0	4	2	8	23
10,000-49,000	0	1	2	0	1	0	1	2	15	22
50,000-99,999	0	0	1	0	0	0	1	2	8	12
100,000+	0	0	0	0	0	0	0	1	1	2
Total	0	5	7	0	2	0	6	7	32	59
Private										
2,000-9,999	3	12	12	2	2	0	0	2	0	33
10,000-49,999	4	15	17	0	1	0	0	1	0	38
50,000-99,999	1	0	5	2	0	0	0	2	1	11
100,000+	1	0	1	2	0	0	0	1	0	5
Total	9	27	35	6	3	0	0	6	1	87

See Appendix A for region code and list of states in each region.

Cities With Franchised Firms

As is shown in Table C-1, 87 cities with franchises were selected for the economic analysis. The private firms and private householders were surveyed in these cities. The firms' managements were given the CON, while householders were telephoned and asked to answer questions on the "Household Survey Questionnaire." Hence, the fee structures provided by private firms were confirmed by householders.

Table C-2 shows that firms in 59 of these cities responded to the CON, with the distribution shown in Table C-3. However, three of these did not supply all the necessary cost data.

Cities with a Private Arrangement

Since many cities with private refuse collection arrangements have more than one refuse collection firm, we had to decide which firm to telephone in each city selected for the sample. This was done by calling residents and asking them the questions on the householder survey and what firm they received service from. Once five households were reached with the same service provider, that firm was called to fill out the CON. Using this method we tried to reach the largest firm in each city. In practice, several large firms were often located and if the main one refused to answer questions, the next largest firm was telephoned. Table C-2 shows that 97 cities with private arrangements received the householder survey and 87 firms that contracted to individual householders responded to the CON survey. These 87 cities were included in the economic analysis. Their population and geographic characteristics are summarized in Table C-3.

Survey Instruments Used to Collect Price Information

The Survey of Municipal Contract Procedures

The survey instrument was sent by mail to city managers or mayors of selected cities. Two mailings were conducted to ensure higher response. These mailings were conducted during March and April 1975. Cities were asked to send contract, ordinance, and bidding documents with the filled-out questionnaire. They were also asked to fill out a second questionnaire concerning local regulation of solid waste collection which was sent to them in the same mailing.

Response Rate and Quality. The response rate after two mailings and follow-up phone calls was 68 percent. Eighty-five cities returned the forms. Because exact contract price information was deemed of high importance, information on 14 additional cities was obtained by telephone. Thus, with the addition of the 14 cities, information was obtained from 99 local jurisdictions, or 79 percent of the sample.

The quality of response was generally good, particularly with regard to the questions on contracting procedures and satisfaction with the contract. The main problem arose with the reported contract price information. It was sometimes not clear what the price included (yard trash, commercial pickup, and so on) despite the fact that cities were asked to give residential refuse contract prices only. Questionable responses were clarified by follow-up telephone discussions with the municipality.

The Private Contractor Survey Questionnaire

This survey was directed to private carting firms. It was used to collect specifications of service provided by the firms and descriptions of the firms' management practices. For all except the cities with municipal contracts, this form was used to record the price schedule offered by the firm to the public.

Logistics of Data Collection. All surveying was done by telephone during the months of March and April 1975. Telephone interviews were completed for 92 contract cities, 59 franchise cities, and 87 nonfranchise cities. An average interview took about 17 minutes, once the appropriate person was reached. A great deal of time was spent making call-backs in order to reach someone who was knowledgeable enough to answer all the questions. Each interviewing shift lasted a maximum of four hours. Five weeks were required to complete the interviews.

The same interviewers who worked on the household survey were used in the contractor survey. Additional training sessions were required to fully prepare the interviewers for the more complex contractor interviews. The extremely low refusal rate (10.6 percent) indicates that interviewing was highly effective. In addition, refusals were avoided by careful institutional identification, by explanations of the general purposes and expected research value of the study, by references to the *Solid Waste Management* magazine article describing the project, and by mentioning the names of individual members of the project advisory panel.

Rigorous procedures were instituted to ensure the quality of the data. All questionnaires were first reviewed by the interviewers themselves and checked for consistency. Then an independent reviewer repeated the consistency checks for each questionnaire. All surveys which did not meet the consistency and accuracy requirements were given back to the interviewers for recalls. After completion of the checking procedures, the surveys were keypunched.

Because of the quality control imposed, responses, for the most part, were reliable. The main problem remained for the tonnage, cubic yard, and load figures reported by the contractors. Even after considerable probing by interviewers, a number of contractors did not know the total amount of residential refuse collected for a specific city. This question presented the main problem for the interviewers and resulted in the elimination of some cities from the phase.

The Householder Survey

The purpose of the household survey was to obtain information from residents of communities who make their own arrangements for refuse collection service. The survey instrument appears at the end of this appendix.

Logistics of Data Collection. Telephone interviews were completed with 1,090 households. An average interview took approximately five minutes.

Experienced interviewers were used, and each was trained specifically for this project. Two-day training sessions included familiarization with the questionnaire, definition of terms, review of the purpose of the survey, role-playing, and actual practice interviews.

Interviewing was done on a staggered-time basis. To minimize fatigue and errors, the maximum length of an interviewing shift was six hours, with appropriate breaks. About four weeks were required to complete the household interviews.

Within the sample cities, individual households were selected from telephone directories. Numbers were randomly drawn from telephone directories as follows:

1. Choose a page number from a table of random numbers.
2. Turn to a selected page and record first phone number of first column. If not a residential number, continue down column to first residential number.
3. Repeat step 2 until required number of households is reached.

Twenty-five residential phone numbers were selected for each sample city. Additional phone numbers were randomly selected when the initial 25 did not produce the expected number of completed interviews.

The object of the householder interviews was to obtain a contractor name, service level rendered, and price charged. In order to check the information obtained, householders were called until 5 householders were reached that used the same contractor. When this occurred, no further calling in that city was done. If interviewing yielded 10 households with similar service levels and fees for refuse collection, although serviced by different firms, that city and the various contractors were included in the sample.

For each city in various population classes a maximum number of calls were made in order to attempt to obtain 5 householders using the same contractor. There were 25 calls for cities from 2,500 to 9,999 population, 50 calls for cities of 10,000 to 49,999, 75 calls for cities for 50,000 to 99,999, and 100 calls for cities over 100,000. If five such householders were not found then the city was excluded from the sample.

A number of other conditions was prescribed whereby the interview was terminated. If it was discovered that less than 65 percent of the city was under a private arrangement, the city was eliminated from the sample. Further, a household interview was terminated if: (1) the household did not arrange for its refuse collection; (2) yard trash was collected separately as an extra service but was not billed separately; (3) the fee paid was unknown; and (4) the householder did not know if there were other arrangements for refuse collection that he could use.

The required number of interviews was completed for 97 private cities and for 65 franchise cities. One thousand and ninety individual household interviews were successfully completed (317 in franchise cities and 773 in nonfranchise cities). The refusal rate for this part of the survey was 14 percent, for both the franchise and private city interviews.

Response Quality. The survey asked interviewers to rate the knowledgeability of the respondent on a scale of 1 to 5, 1 representing the highest degree of knowledgeability and 5 the lowest degree. Of the 1,089 interviews conducted, interviewers ranked 49 percent in the most accurate category, 37 percent in the next category, 11 percent in the fairly accurate category, and the remaining 3 percent in the least accurate category.

The quality of response varied from question to question. Interviewers were trained to ensure that the price information that was given was accurate. In addition, interviewers tried to make sure that the respondent gave the contractor's name and address. The least trustworthy responses were those given to the questions about householder choice of service provider and the number of other firms operating in the neighborhood. Responses to the question of whether the private firm puts limits on householders as to the amount of garbage it will pick up were also difficult to interpret. One could not interpret whether the householders' responses reflected official policy or informal practice of the service provider.

SURVEY OF MUNICIPAL CONTRACT PROCEDURES FOR

RESIDENTIAL SOLID WASTE COLLECTION

COLUMBIA UNIVERSITY
Government Performance Study Group
SOLID WASTE MANAGEMENT PROJECT

DEFINITION: SOLID WASTE -- Discarded material consisting of garbage, trash or a combination thereof.

A. CURRENT STATUS OF CONTRACT(S)

1. How many contracts does your municipality currently have in effect for residential solid waste collection (e.g., regular house-to-house collection exclusive of any special contracts for yard trash, bulk, etc.).... _____ 15

2. How many <u>firms</u> are currently under contract to your municipality to collect residential solid waste? .... _____ 16-17

3. If your contract(s) specifies a lump sum dollar amount to be paid to contractors, what is the basic lump sum which you are paying? (PLEASE ANSWER FOR EACH CONTRACT YOU MAY HAVE.) ....

   a. Contract # 1 .... $_____. 18-23

   b. Contract # 2 .... $_____. 24-29

   c. Contract # 3 .... $_____. 30-35

4. Does your contract(s) cover all households in the city?.. YES ( ) NO ( ) 36

   If "NO", approximately what % of households are covered?.. _____ %
   37-38

5. Does the municipality bill residents for solid waste collection services exclusive of any special charges for yard trash, bulk pickups, etc. .... YES ( ) NO ( ) 39

   If "YES", what is the current basic fee? ....$_____ per _____ 44
   40-43   (Month, year, quarter)

B. CONTRACTING PROCEDURES

6. How is the contractor selected? (PLEASE CHECK CHOICE BELOW.) ....

   __45__ a. By open competitive bidding procedures without subsequent negotiation of terms

   __46__ b. By negotiation (without bidding)

B. 6. ⁴⁷____ c. By a combination of above (e.g., contractor must submit a competitive bid. Specific services to be rendered and rates are then negotiated upon award of contract.)

⁴⁸____ d. Other (PLEASE EXPLAIN) _____

_____

7. How are bids solicited? (PLEASE CHECK CHOICE BELOW)

⁴⁹____ a. By general public announcement
⁵⁰____ b. By specific invitation
⁵¹____ c. Both
⁵²____ d. Other (PLEASE EXPLAIN) _____

_____

8. By what criteria is the contract(s) awarded? (PLEASE CHECK CHOICE BELOW.)

⁵³____ a. Contract must go to lowest bidder
⁵⁴____ b. Contract goes to lowest <u>qualified</u> bidder
⁵⁵____ c. Other (PLEASE EXPLAIN.) _____

_____

9. Is the public given the opportunity to challenge or contest the contract award .... YES ( ) NO ( )₅₆

10. For your present contract(s), what were the total number of bidders for each?

    a. Contract # 1 .... _____ number of bidders. ⁵⁷⁻⁵⁹
    b. Contract # 2 .... _____ number of bidders. ⁶⁰⁻⁶²
    c. Contract # 3 .... _____ number of bidders. ⁶³⁻⁶⁵

C. CONTRACT TERMS

11. Has the service area for residential solid waste collection to be covered by the current contract(s) increased (for example, through annexation) or decreased during the present contract period? .... YES ( ) NO ( )₆₆

12. Is approval of the municipality required for changes in the contractor's pick-up schedules? .... YES ( ) NO ( )₆₇

13. Have any provisions in your current contract(s) (other than chose for rates, routes, or service area) been altered or amended during the life of the current contract? .... YES ( ) NO ( )₆₈

C. 13. If "YES",
   a. Who initiated the change?
      _69_ 1. Municipality
      _70_ 2. Contractor
   b. Was the change
      _71_ 1. Negotiated
      _72_ 2. Unilateral

   _73_ c. What terms were changed? _____
   _____

14. How many employees do you have which do the following tasks connected with contracting? (Please include fraction of employees' time which may be spent on an annual basis on one of the tasks below.)

	Number of Employees
a. Monitoring performance of the contractor	_____ 74-76
b. Handling customer complaints	_____ 77-79
c. Preparation and sending of bills to customers ...	_____ 15-16

15. Does the municipality provide the contractor(s) with a municipally owned or operated disposal site for the disposal of refuse? .... YES ( ) NO ( )17

   If "YES", Does the municipality charge the contractor(s) a fee for use of the site? .... YES ( ) NO ( )18

D. SATISFACTION WITH CONTRACT

	YES	NO	DON'T KNOW
16. Is (Are) your present contract(s) a renewal of a previous contract with the same firm(s)?	( )	( )	( )19
a. Contract # 1	( )	( )	( )20
b. Contract # 2	( )	( )	( )21
c. Contract # 3	( )	( )	( )22
17. Do any of your contracts require a performance bond?	( )	( )	( )23
If "YES", has such a bond been forfeited to the municipality, to your knowledge?	( )	( )	( )24

D. 18. Has any breach of performance occurred under your present contract(s)? .... YES( ) NO( )$_{25}$

19. Will you probably renew your present contract(s) with the same firm(s) when it (they) expire? .... YES( ) NO( )$_{26}$

    If NO, why not? _____
    _____

20. Are you likely to be using contracts for residential collection five years from now? .... YES( ) NO( )$_{27}$

COULD YOU KINDLY SEND US:

(1) YOUR <u>MOST RECENT</u> CONTRACT(S).

(2) COPIES OF YOUR ORDINANCES CONCERNING SOLID WASTE COLLECTION.

(3) COPIES OF YOUR ORDINANCES GOVERNING BIDDING PROCEDURES FOR PRIVATE HAULERS PROVIDING RESIDENTIAL SOLID WASTE COLLECTION.

(4) A COPY OF YOUR BIDDING INSTRUCTIONS, SPECIFICATIONS, AND PROCEDURES.

(5) A COPY OF YOUR ANNOUNCEMENT/INVITATION TO BIDDERS IF DIFFERENT FROM (4).

THANK YOU FOR YOUR ASSISTANCE. PLEASE FILL IN:

Your Name: _____ Telephone # _____

Title: _____ Date _____

Name of Municipality _____

PLEASE RETURN TO:

COLUMBIA UNIVERSITY
SOLID WASTE MANAGEMENT PROJECT
608 URIS HALL
NEW YORK, N. Y. 10027

ENVELOPE IS ENCLOSED

Columbia University

PRIVATE CONTRACTOR SURVEY QUESTIONNAIRE

Solid Waste Management Project

SMSA Code ⊔⊔⊔⊔ Phone No. (⊔⊔⊔)⊔⊔⊔⊔⊔⊔⊔
Place Code ⊔⊔⊔⊔ Interviewer ⊔⊔
County Code ⊔⊔⊔ Type of City   ⊔
MCD Code ⊔⊔⊔   1 = franchise
                     2 = non-franchise
Interviewee Code ⊔⊔ 3 = city contract
Date ⊔⊔ ⊔⊔ |1|9|7|5|
     month day   City Size
                    1. 2,500 - 9,999   ⊔
SMSA Name _____ 2. 10,000 - 24,999
                    3. 25,000 - 49,999
City Name _____ 4. 50,000 - 99,999
                    5. 100,000 or more

   Region
1. New England   6. East South Central
2. Middle Atlantic   7. West South Central   ⊔
3. East North Central   8. Mountain
4. West North Central   9. Pacific
5. South Atlantic

INTERVIEWER: ESTIMATE THE OVERALL ACCURACY OF RESPONDENT'S ANSWERS:

      Very   Not
    Accurate   Accurate
       1   2   3   4   5   ⊔

[Hello, I'm ___(name)___ and I'm with the Business School at
Columbia University in New York City. I'm phoning you as a part of
a national study of solid waste collection and disposal services.
I'd like to ask some general questions about the services provided
by ___name of firm___] Can I speak with the Office Manager or
person in charge of Public Relations, please?
     REPEAT PART IN BRACKETS WHEN CONNECTED, OR OBTAIN NAME OF
PERSON TO BE CONTACTED AND TIME TO CALL BACK.

     We are phoning about 300 firms who provide refuse collec-
tion services - they're located throughout the United States.
We plan to average the responses to the questions I'll ask, so
our final report will contain accurate information about the
industry, but no data on any individual firm.
     In the first stage of our project, we phoned 1400 small and
medium-sized cities to ask who picked up the refuse. We under-
stand from that survey that you provide residential refuse collection services

1. in ___(name)___ city. Is this correct? 1 = yes, 2 = no |__|
                                                             37
   IF YES, GO TO Q 1c

   a) IF NO, Do you know who does collect refuse there?
      IF A PRIVATE FIRM, OBTAIN NAME AND ADDRESS AND TEL. NO.

      NAME: _____

      ADDRESS: _____

      PHONE: _____

   b) IF NO, Did you at one time provide collection services
      to ___(name)___ city? 1 = yes, 2 = no |__|
                                                              38
      When did you stop collecting there? |__|__| 197 |__|
                                              month   41
                                              39 40

      IF THE CONTRACTOR COLLECTED DURING 1974, CONTINUE THE
      INTERVIEW, OTHERWISE, TERMINATE INTERVIEW.

   c) Do you also provide commercial and/or industrial
      pick-up service in ___(city)___? 1 = yes, 2 = no |__|
                                                              42

   Please give me information that pertains only to residential
   collection operations in ___(city)___.

   First, I'd like to ask you:

2. For how many years have you provided residential collec-
   tion services in ___(city)___?   # years |__|__|
                                                           43 44

3. Do you provide regular (garbage and trash) service to all
   the households in ___(city)___? (INTERVIEWER: MAKE SURE
   ANSWER DOES NOT INCLUDE RURAL COLLECTION BOXES)
        1 = yes, 2 = no |__|
                                                            45
   IF NO:
   a) For how many households do you provide regular service in
      ___(city)___?
                                              |__|__|__|__|__|__|
                                               46 47 48 49 50 51

4. a) What % is that of all the households in the city? |__|__|__| %
                                                         52 53 54

   b) What percent of the households you service would you
      say live in:

             single-family houses? |__|__|__| %
                                              55 56 57

             multi-family housing? |__|__|__| %
                                              58 59 60

c) Do you operate in one area of the city, or in various areas scattered throughout the city?

    1 = one area

    2 = various scattered areas   ☐
                                                                                                 61

5. If other firms also pick up residential refuse in __(city)__, can you estimate how many other firms also service __(city)__?

                                                                        # of firms ☐☐
                                                                                                     62 63

6. How frequently do you provide service? Can you tell me: What percent of the households for which you provide regular service do you service.... (ADD SEPARATE TRASH AND WET GARBAGE COLLECTIONS TOGETHER, EXCLUDE SEPARATE YARD TRASH COLLECTION)

    a. once a week   ☐☐☐ %
                                                                                     64 65 66

    b. twice a week   ☐☐☐ %
                                                                                     67 68 69

    c. "other" times per week (SPECIFY #) ☐☐   ☐☐☐ %
                                                         70 71   72 73 74

    d. other times (SPECIFY) _____   ☐☐   ☐☐☐ %
                                                             75 76   77 78 79

                                                                       CARD I   1
                                                                                                   80

7. Where do you pick up refuse? From what percent of the households for which you provide regular service do you pick up at....

    a. curbside   ☐☐☐ %
                                                                                    17 18 19

    b. backyard   ☐☐☐ %
                                                                                    20 21 22

        1 = can returned   3 = can not ret'd

        2 = "barrel" system   4 = "other" _____   ☐
                                                                                             23

    c. alley   ☐☐☐ %
                                                                                    24 25 26

    d. "other (SPECIFY) _____   ☐☐☐ %
                                                                                    27 28 29

8. Is there an upper limit on the # of cans that you will collect from any one household? 1 = yes, 2 = no   ☐
                                                                                                 30

IF YES, what is this limit?   # of cans ☐☐
                                                                                       31 32

9. What percent of household refuse is collected in....

        1. returnable containers (cans)   ☐☐☐ %
                                                                             33 34 35

        2. non-returnable containers (bags)   ☐☐☐ %
                                                                             36 37 38

10. Can you tell me how much total refuse you collect from your residential accounts in ___(city)___ the last year?

    a. tons per year
       1 = estimated
       2 = weighed

    b. cubic yards per year
       1 = estimated
       2 = measured
       Is that compacted?
       1 = compacted
       2 = not compacted

    c. loads per year (cu. yds/load _____)
       1 = estimated
       2 = counted

11. Are there any costs to your firm for disposing of residential refuse? 1 = yes 2 = no

    IF 2 ON Q11, GO TO 14, OTHERWISE ASK Q12

12. Is that cost:
       1 = a charge for disposing residential refuse?
              OR
       2 = some other kinds of costs?

    IF 1 ASK Q12a
    IF 2 ASK Q12b

    a. What are you charged for disposing of residential refuse? $
       Is that charge....
       1 = per load
       2 = per cubic yard
       3 = per ton
       4 = a total annual charge
       5 = a total monthly charge
       6 = other (SPECIFY) _____

    b. What kinds of costs were those? (INCLUDE ALL THAT APPLY)
       1 = site ownership or rental
       2 = site operation and maintenance
       3 = other (SPECIFY) _____

13. What were your total residential refuse disposal costs for last year in ___(city)___ ? $

    CARD II

14. Are your residential collection rates set by the city or state? 1 = yes, 2 = no

15. How do you charge for your regular residential service?
    Is it a....
        1 = flat rate
        2 = variable rate, i.e., depending
            on the service |_|
                                                     18
16. What is your rate schedule? (RATE PER HOUSEHOLD)

Option	$	$ per	Content	Point of pickup	Frequency of pickup	Quantity	
A	|_|_|_|_|_| 19 20 21 22 23	|_| 24	|_| 25	|_| 26	|_| 27	|_| 28	
B	|_|_|_|_|_| 29 30 31 32 33	|_| 34	|_| 35	|_| 36	|_| 37	|_| 38	
C	|_|_|_|_|_| 39 40 41 42 43	|_| 44	|_| 45	|_| 46	|_| 47	|_| 48	
D	|_|_|_|_|_| 49 50 51 52 53	|_| 54	|_| 55	|_| 56	|_| 57	|_| 58	
E	|_|_|_|_|_| 59 60 61 62 63	|_| 64	|_| 65	|_| 66	|_| 67	|_| 68	CARD III
F	|_|_|_|_|_| 69 70 71 72 73	|_| 74	|_| 75	|_| 76	|_| 77	|_| 78	|3| 80
G	|_|_|_|_|_| 17 18 19 20 21	|_| 22	|_| 23	|_| 24	|_| 25	|_| 26	
H	|_|_|_|_|_| 27 28 29 30 31	|_| 32	|_| 33	|_| 34	|_| 35	|_| 36	

Codes:
```
            1 = week       1 = garbage   1 = curb   1 = 1 of    1 = per cu yd
            2 = month          only      2 = front      7 days      (SPECIFY
            3 = quarter    2 = trash         yd.    2 = 2 of    _____)
            4 = year           only      3 = back       7 days  2 = per can
            5 = other      3 = garbage       yd.    3 = 3 of       (SPECIFY
                (SPECIFY)      & trash    4 = alley      7 days  _____)
                           4 = bulk                  4 = 4 of    3 = per bag
                           5 = yard                      7 days     (SPECIFY
                               trash                  5 = 5 of   _____)
                                                         7 days  4 = no limit
                                                      6 = 6 of   5 = per other
                                                         7 days     (SPECIFY
                                                      7 = 7 of   _____)
                                                         7 days
```

For Comments:

Now I'd like to ask just a few more questions about your employees and the equipment you use.

First, about your employees:

17. How many drivers and collectors work full time (or their equivalent) in collecting household refuse in __(city)__?

18. On the average, what does a collector earn per month, excluding fringe benefits? $

19. How many hours are there in a normal work week?

20. How many hours are there in a crew shift?

21. How many days per week does collection of household refuse in __(city)__ take place?

22. How many collection crews work in __(city)__ on an average collection day?

23. On the average, how many vacation days do your collectors get per year?

24. What would you estimate your average absentee rate to be? That is, what percent of your work force is absent on an average day? (THIS INCLUDES ALL ABSENCES--FOR ILLNESS, VACATION, OR INJURY). %

25. Are collection crews on an incentive system?
    1 = yes 2 = no

    a) IF YES, will you describe this system?

    _____
    _____
    _____

26. Are any of your residential collection employees unionized? 1 = yes, 2 = no

    a) IF YES, which union(s) do they belong to? (CODE ALL THAT ARE APPLICABLE)

        1. Local Teamster

        2. Other (SPECIFY) _____

27. Are you a subsidiary of a national firm?
    1 = yes   2 = no   9 = don't know   ⊔
                                           68

    IF YES, What is the name of your parent firm?

    Name: _____

Now, about your equipment

28. How many of your firm's trucks pick up residential waste
    on an average work day in __(city)__ ? # of trucks ⊔⊔⊔
                                                        69 70 71

    INTERVIEWER: IF THE FIRM COLLECTS IN OTHER CITIES (OR AREAS)
                 AS WELL, ASK THE RESPONDENT TO ESTIMATE HOW
                 MANY TRUCKS THE FIRM WOULD USE FULL TIME (i.e.,
                 the equivalent # of trucks working five days a
                 week) TO PICK UP THE GARBAGE FROM THEIR ACCOUNTS
                 IN __(city)__ .

    a) INTERVIEWER: INDICATE WHETHER THE ABOVE NUMBER IS AN
                    ESTIMATE OR ACTUAL COUNT.

        1 = estimate   ⊔
                                                        72
        2 = actual count

29. How many truck shifts (for residential collection)
    operate in an average week in __(city)__ ?
                                         # of shifts ⊔⊔⊔
                                                     73 74 75

    a) Is the number of paid crew shifts per week different
       from the number of truck shifts? If so, how many crew
       shifts?   # shifts ⊔
                                                         76

30. What types of trucks does your firm use for residential
    collection in __(city)__ ? And how many of each type do
    you have?
                                          CARD IV |4|
                                                   80

	# owned	# used in city	crew size	cu.yd. cap.	# comp.
Rear loader	⊔⊔⊔ 17 18 19	⊔⊔⊔ 20 21 22	⊔⊔ 23 24	⊔⊔ 25 26	⊔⊔⊔ 27 28 29
Front loader	⊔⊔⊔ 30 31 32	⊔⊔⊔ 33 34 35	⊔⊔ 36 37	⊔⊔ 38 39	⊔⊔⊔ 40 41 42
Side loader	⊔⊔⊔ 43 44 45	⊔⊔⊔ 46 47 48	⊔⊔ 49 50	⊔⊔ 51 52	⊔⊔⊔ 53 54 55
Other (SPECIFY) _____	⊔⊔⊔ 56 57 58	⊔⊔⊔ 59 60 61	⊔⊔ 62 63	⊔⊔ 64 65	⊔⊔⊔ 66 67 68

31. How long (in minutes) does it take a truck, on average, to leave its route in __(city)__, drive to the disposal site, dump, and return to the route? minutes |__|__|__|
 69 70 71
32. What percent of residential collection trucks are used in __(city)__ dump at landfills? |__|__|__|%
 72 73 74
 CARD V |5|
 80
33. What percent of households in __(city)__ would you say are serviced by:
 a. private firms? |__|__|__|%
 17 18 19
 b. municipality? |__|__|__|%
 20 21 22
 c. self-service? |__|__|__|%
 23 24 25
34. Do you have a contract with the city for residential collection in __(city)__ ? 1 = yes, 2 = no |__|
 26

 IF 1 on Q34 GO TO Q36

35. Did your firm have to be franchised before you could collect in __(city)__ ? 1 = yes, 2 = no |__|
 27

 IF FIRM NOT UNDER CONTRACT WITH CITY, TERMINATE INTERVIEW.

36. Did your firm have to bid competitively for this contract? 1 = yes, 2 = no |__|
 28

 IF 2 on Q36, GO TO Q39

37. How many other firms bid for this contract? # of firms |__|__|
 29 30
38. Why did your firm get the contract?
 1 = lowest bidder
 2 = other reasons (SPECIFY) _____ |__|
 31
39. Are you satisfied with the terms of the contract? |__|
 1 = satisfied as it is 32
 2 = satisfied, although it could be better
 3 = dissatisfied with it as it is now
 9 = don't know / no answer

 IF 2 OR 3 ON Q39, ASK Q40. OTHERWISE TERMINATE

40. How could the terms of this contract be improved? |__|__|
 33 34
_____
_____
_____
 CARD VI |6|
 80

Columbia University

HOUSEHOLD SURVEY QUESTIONNAIRE

Solid Waste Management Project

SMSA Code ☐☐☐☐ Phone No. (☐☐☐) ☐☐☐ ☐☐☐☐
Place Code ☐☐☐☐ Interviewer ☐☐
County Code ☐☐☐ Type of City ☐
MCD Code ☐☐☐   1 = franchise
Interviewee Code ☐☐   2 = non-franchise
Date ☐☐ ☐☐ |1|9|7|5|
     month day

SMSA Name _____

City Name _____

Region ☐
1. New England
2. Middle Atlantic
3. East North Central
4. West North Central
5. South Atlantic
6. East South Central
7. West South Central
8. Mountain
9. Pacific

City Size ☐
1. 2,500 - 9,999
2. 10,000 - 24,999
3. 25,000 - 49,999
4. 50,000 - 99,999
5. 100,000 or more

INTERVIEWER: ESTIMATE THE OVERALL ACCURACY OF RESPONDENT'S ANSWERS:

Very Accurate			Not Accurate	
1	2	3	4	5

☐

Hello, I'm ___(name)___ and I am with the Business School at Columbia University in New York. We are doing a national study of how Americans get their garbage collected and how they feel about their garbage service. We hope that the information we get from you and others in our nation-wide sample will help us come to some scientific conclusions about how garbage service can be improved. You would help us a great deal if you would take two or three minutes to answer a few questions about your garbage service. First, I would like to ask you:

1. Who collects your household garbage? By household garbage, I mean things like used Kleenex, tomato soup cans, etc.
    - 0 = you, yourself
    - 1 = a dept. of your city govt.
    - 2 = a private firm under contract with your city or county
    - 3 = a private firm that you hire yourself
    - 4 = the county
    - 5 = another city
    - 6 = a special authority
    - 7 = a voluntary association
    - 8 = other (SPECIFY) _____
    - 9 = don't know (Is there someone in the house who would know? IF NO, TERMINATE)

    ⊔ 37

    IF ANSWER TO Q1 IS 3, GO TO Q2. OTHERWISE GO TO Q 13

2. Are any of the following collected separately from wet garbage (food wastes)?
    1 = yes, 2 = no, 9 = don't know

    a. dry trash ⊔ 38

    b. yard trash ⊔ 39

    IF YARD TRASH COLLECTED SEPARATELY, ASK:

    c. is yard trash billed separately from household garbage?
       1 = yes, 2 = no, 9 = don't know ⊔ 40

    IF 2b = 1 and 2c = 2, GO TO Q13

3. What do you pay for garbage service? $ ⊔⊔⊔⊔⊔ 41 42 43 44 45

4. Is that charge per....
    - 1 = week
    - 2 = month
    - 3 = quarter
    - 4 = year
    - 5 = other (SPECIFY _____)
    - 9 = don't know

    ⊔ 46

    IF 9 on Q4, GO TO Q13.
    SEE CITY TERMINATION PROCEDURE

5. How many times a week is your garbage picked up? [IF WET AND DRY HOUSEHOLD REFUSE PICKED UP SEPARATELY, ADD TOGETHER]

    Times per week............................... ⊔ 47

6. Where is your garbage picked up? .... ☐
    1 = curbside   48
    2 = backyard, can returned
    3 = backyard, can not returned
    4 = backyard, barrel system
    5 = alley
    6 = other (SPECIFY) _____

7. Is there any limit on the amount of garbage picked up? ☐
    1 = yes, (GO TO Q7a)   49
    2 = no   (GO TO Q8)
    3 = don't know (GO TO Q8)

    a. IF YES, ASK: How is the garbage that is over
       the limit collected? .... ☐
       1 = hauler collects with extra charge   50
       2 = household carts away
       3 = other (SPECIFY) _____
       4 = don't know

8. Would you please tell me the name of your garbage collection
   firm? What is the address? What is the phone number?
        Name: _____

        Address: _____

        Phone: ( ) _ _ _ - _ _ _ _

9. Do you know if any other firms collect in your neighborhood?
    1 = yes   2 = no   9 = don't know   ☐
                                                          51

    IF 1 on Q9, ASK:
10. How many other firms?   ☐☐
    1 = one   52 53
    2 = two
        etc.

11. Could you have your garbage picked up by another firm if
    you wanted, or is your present firm your only option?
    1 = other options   ☐
    2 = no other options   54
    9 = don't know

    IF 1 on Q11, ASK Q12. OTHERWISE GO TO Q13.

12. What are the other ways you could have your garbage collected?
    Could you.... CODE ALL OPTIONS
    0 = cart it away yourself
    1 = have your city pick up   ☐
    2 = have a private firm under contract   55
        with the city pick up   ☐
    3 = have another private firm hired by you pick up   56
    4 = have the county pick up
    5 = have another city pick up   ☐
    6 = have a special authority pick up   57
    7 = have a voluntary association pick up
    8 = other (SPECIFY) _____   ☐
    9 = don't know   58

13. How satisfied are you with your present garbage collection? Would you say you are.... ⎵ 59
    1 = satisfied as it is (TERMINATE INTERVIEW)
    2 = satisfied but it could be better (GO TO Q14)
    3 = dissatisfied with it as it is now (GO TO Q14)
    9 = don't know (TERMINATE INTERVIEW)

14. How do you think your garbage collection could be improved? By.... (1 = yes, 2 = no, 9 = don't know)

    collecting more frequently..................... ⎵ 60

    charging less.................................. ⎵ 61

    picking up at a different place _____ ⎵ 62

    changing the amount collected.................. ⎵ 63

    a reduction in the noise of pickup............. ⎵ 64

    a reduction in spilled garbage................. ⎵ 65

    a reduction in missed collections.............. ⎵ 66

    garbage collection more on time................ ⎵ 67

    collectors more courteous...................... ⎵ 68

    those responsible more responsive to complaints ⎵ 69

    other (SPECIFY) _____ ⎵ 70

Index

Index

Advisory Commission on Intergovernmental Relations, 2, 169
Allegheny, Pa., 22
American Public Works Association, 3, 28, 37-38, 41, 44, 47, 52, 54-55
Amiens, France, 16
Applied Management Sciences Survey of *1971,* 39-40, 44, 47
Arizona, 179, 180
Arkansas, 179, 181
Arrangement, multiple service, 31-32, 39, 41-42, 49, 55
Arrangements, service. *See* Contract collection; Franchise collection; Municipal collection; Private collection; Self-service collection
Athens, Greece, 13
Avignon, France, 15

Baltimore, Md., 21, 107
Bentham, Jeremy, 20
Bidding, competitive, 140, 147, 151n.13, 154, 165
Billing, 86-89, 91, 122, 134, 140, 147-148
Blegen, Carl William, 12-13
Board of Health (Boston), 18-19
Bologna, Italy, 15
Boston, Mass., 18-19, 21, 100, 107, 108, 109, 131
Bristol, England, 19
Brooklyn, N.Y., 142
Bureau of Solid Waste Management of the U.S. Public Health Service (PHS), 38
Byzantium, 14

Calcutta, India, 22
California, 2, 108, 126, 180
Charles Town, N.C., 19
Chicago, Ill., 22, 109
China, 14
City Engineering Data Survey, 40-41

Civil Service, 123-124
Clark, Robert M., 108
Cleveland, Ohio, 21
Colorado, 176
Columbia University, 44, 67
Commission of Sewers (England), 16
Congress (U.S.), 170, 173
Connecticut, 180, 181
Contract collection, 29, 32, 37, 38, 41, 52, 64, 139-140
 cost of, 124-125, 127-128, 131, 134-135, 136, 136n.1, 137n.14, 137n.15, 199-200 (*see also* Cost components of)
 vs. francise collection, 146-149
 vs. municipal collection, 127-128, 133-135, 136, 200, 201-202
 public regulation and, 123, 125
Contract document, between municipality and private firm, 153-157, 159, 161, 163-165, 167-168
Cost, components of, *See also* Franchise collection, cost of; Municipal collection, cost of; Private collection, cost of; Self-service collection
 absenteeism, 110, 111, 114, 117
 capital equipment, 108-109, 116
 capital intensity, 110, 111, 117
 crew size, 110, 111, 113-114, 116-117
 fringe benefits, 100, 103, 106-107
 incentive system, 110, 111, 113, 114, 117
 management, 110-115, 116-117
 operation expenses, 100, 103, 104, 106
 overhead, 100, 103, 104, 107
 service levels, 108-109, 110
 vehicle, 100, 103, 110, 111, 113-114, 116, 117
 wages, 99-100, 101-102, 103-104, 106, 110, 116, 130

Cost of collection, 127-136, 136n.1
  management and, 110-115, 116-117, 129-130, 131, 136n.3., 137n.14
Council of Athens (Greece), 13
Council of State Governments, 174
Crete, 12
Crime, organized, 144

Deccan, 12
Delaware River, 19
Denver, Colo., 21
Department of the Interior (U.S.), 174
Department of Labor (U.S.), 172
Department of Transportation (U.S.), 172
District of Columbia, 38, 179
Dublin, Ireland, 19
Duffy, John, 21
Dumps, refuse, 4, 12-22 *passim*, 183

Economy
  of contiguity, 140, 143, 148, 149, 150n.5
  of scale, 125-126, 131-132, 142-143, 148, 149, 202
Edinburgh, Scotland, 19
Edward the Third (king of England), 15
Employees, public, 1, 2, 123-124, 125, 136n.3
Energy Supply and Environmental Act of *1974*, 173
England, 16, 17, 19, 20, 21
Environment, 169, 170, 176, 177
EPA (Environmental Protection Agency), 171, 173

Federal Aid Highway Act of *1956*, 172
Federal Clean Air Act, 169, 173
Federal Noise Control Act, 172
Federal Solid Waste Disposal Act of *1965*, 169, 170-171, 173, 174, 176, 182
Fees,
  flat, 79-80, 85-86

  variable, 79-80, 84, 85-86, 96
Florence, Italy, 15
Franchise collection, 29, 38, 55, 64, 67, 73, 84-85, 140, 175, 180
  and billing, 87-88
  vs. contract collection, 146-149
  cost of, 124-125, 127-129, 131, 133-134, 135, 136, 136n.1, 136n.2, 137n.4, 137n.7 (*see also* Cost, components of)

Germany, 15-16
Gillian, James I., 108
Glasgow, Scotland, 20
Government, county, 26-27
Government, municipal,
  and civil service, 123-124
  vs. free market, 141-144, 149
  productivity in, 1-2
  as service arranger, 27
  as service provider, 26, 45, 47, 63, 64
  as service recipient, 25-26, 47, 49, 52, 53

Hangchow, China, 14
Hawaii, 181
Health, public, 141-142, 149
Henry the Fourth (king of France), 17
Henry VI (king of England), 16
Henry VII (king of England), 16
Herakopolis, Egypt, 12

Incineration, refuse, 4, 22
India, 12
Indiana, 31-32, 175
Industry, collection from, 4, 25, 38
Institutions, collection from, 8, 25, 38, 63
Internal Revenue Code, 173
International City Management Association and Public Technology, Inc. (ICMA), 41-42, 44, 47
Intervention, government
  vs. free market, 141-144, 149

Jerusalem, 12
Johnson, Lyndon B., 169

Kansas, 175, 180
Kennebunkport, Maine, 19

Labor, 123-124, 131, 172-173
"Lakewood Plan," 2
Liverpool, Wales, 21
London, England, 15, 17
Los Angeles, Calif., 22
Louisiana, 176

McFarland, Jean M., 126
Mahenjo-Daro, 11
Management, solid waste, 170-173
    and state legislation, 173-183
Manhattan, New York, N.Y., 17-18, 21
Maryland, 21, 107
Massachusetts, 18-19, 21, 100, 107, 108, 109, 131
MIT (Massachusetts Institute of Technology), 22, 35
Mauryan empire, 12
Memphis, Tenn., 20
Mesoamerican civilizations, 17
Michigan, 175, 180, 182
Milan, Italy, 15
Milwaukee County, Wis., 175
Minnesota, 2, 21, 175, 179, 182
Mississippi, 182
Missouri, 21, 180
Monopoly, collection, 202
    vs. government intervention, 142-143
Morse, William F., 35
Mosaic Law, 12
Mumford, Lewis, 11, 15
Municipal collection, 29, 32, 35-64 *passim*, 100, 127
    and billing, 86
    in cities, 67, 72-73
    vs. contract collection, 127-128 133-135, 136, 200, 201-202
    cost of, 124, 127-128, 131, 133, 134-135, 136, 136n.1, 136n.3, 137n.15, 199-201 (*see also* Cost, components of)
    and mandatory service, 84
    vs. private collection, 121-123, 136n.1, 136n.2, 136n.3, 137n.14, 137n.15
    variable fees and, 84
Municipal Index Survey of *1929*, 35-36

Naples, Italy, 16
National Center for Resource Recovery, 8
National Commission on Productivity, 2, 97
National Symposium on State Environmental Legislation, 174
National System of Interstate and Defense Highways, 172
Nehemiah, 12
Netherlands, 17
New Orleans, La., 20
New York, 180
New York City, 18, 20-21, 22
Niskanen, William A., 124
North Carolina, 180
North Dakota, 179
Nuremberg, Germany, 15-16

Occupational Safety and Health Act of *1970* (OSHA), 172, 173
Ohio, 177
Oklahoma, 108
Omaha, Nebr., 21
Oregon, 180
Ospitalieri di Sant'Antonio, 15
Oxford, England, 16 ·

Paris, France, 15, 17, 19
Parliament (England), 15
Partridge, Lawrence, 130
Pennsylvania, 179
Peterson, George E., 72
Philadelphia, Pa., 19, 21
Population, 199-200
    economy of scale and, 125-126, 131-132, 202
    and solid waste, 11, 13, 14, 17, 20, 21, 22, 37
Portland, Oreg., 100

Portsmouth, England, 17
Private collection, 26, 29, 32, 35-64 *passim*, 67, 73, 127, 200
  and billing, 87-88
  in cities, 67
  costs of, 124-125, 127-129, 131, 133-134, 135, 136, 136n.1, 136n.2, 137n.7, 137n.14 (*see also* Cost, components of)
  and mandatory service, 84-85
  vs. municipal collection, 121-123, 124-125
  variable fees and, 84
Proclamation of *1372* (England), 15
Productivity, monopoly and, 122-123
Profits, 122
Providers, service. *See also* Contract collection; Franchise collection; Municipal collection; Private collection; Self-service collection
Public Health, 141-142, 149
Public Health Act of *1875* (England), 21
Public Health Service Survey of *1968*, 38-39, 47
Public property, collection for, 4, 25-26, 47, 49, 52, 63
Public Works, 40-41

Recipient, service
  commerce as, 25, 37, 38, 39, 40
  contract collection and, 140
  contractual obligations of, 156
  franchise collection and, 140
  household as, 25, 37, 38, 39, 40
  industry as, 25, 39, 40, 47
  institutions as, 25, 39, 47, 52
  municipal government as, 25-26, 37, 38, 47
  as service arranger, 27
  as service provider, 26, 45, 49, 59, 84
"Record of Institutions," 12
Resource Recovery Act of *1970*, 169, 170-171, 182
Rhode Island, 180
Rome, Italy, 13-14

Roosevelt Island, New York, N.Y., 22
Rouen, France, 16

Safety, occupational, 172-173
Saint Louis, Mo., 21
Saint Omer, France, 15
Saint Paul, Minn., 21
San Francisco, Calif., 100
Sanitary Act of *1845* (England), 21
Scavenging, 11-19 *passim*
Schreiner, Dean, 108, 109
Scotland, 19, 20
Self-service collection, 26, 27, 29, 32, 38-55 *passim*, 84
  diversification and, 124, 125
  exclusivity of, 123, 125
  mandatory, 84-85, 123, 125
  modes of payment for, 79-81, 84-85
  required, 178-179
Service, collection, government intervention vs. free market, 143-144, 149, 150n.6, 150n.7
Settlers, Dutch, 17-18
Siena, Italy, 15
Smith, Frank A., 8
South Carolina, 179, 181
South Dakota, 180
Spain, 16-17
Standard Metropolitan Statistical Areas (SMSAs), 38, 40, 44-45, 59, 61, 63
"State Solid Waste Management and Resource Recovery Incentive Act," 174, 176
Strikes, employee, 143-144
Suggested State Legislation (1972), 174

Talmud, 12
Taxes, 79-80, 81, 84, 85-86, 88-89, 91, 96, 122, 199-200
Technology, refuse, 21, 22, 25, 101, 116
Tokyo, Japan, 17, 21
Troy, 12
Tweed, William Marcy ("Boss"), 20

Ur, 11
U.S. Council on Environmental
    Quality, 174
Utah, 176
Utilitarians, 20

Valencia, Spain, 16-17
Vehicle, collection, 22, 109, 110, 111,
    113-114, 116, 117
Venice, Italy, 15

Waring, George E., 21
Waterways, solid waste in, 4, 12-22
    passim
Webb, Beatrice, 20
Webb, Stanley, 20
Wisconsin, 175
Worcester, Massachusetts, 21
Wyoming, 176

List of Contributors

Daniel Baumol
Research Associate
Center for Government Studies
Graduate School of Business
Columbia University

Franklin R. Edwards
Professor
Graduate School of Business
Columbia University

Frank P. Grad
Professor of Law
Columbia University Law School
Director, Legislative Drafting
  Research Fund

Bennett C. Jaffee
School of Law
Columbia University

Christopher Niemczewski
Research Assistant
Center for Government Studies
Graduate School of Business
Columbia University

Barbara J. Stevens
Assistant Professor
Graduate School of Business
Columbia University

William Wells
Assistant Professor
Graduate School of Business
Columbia University

About the Author

E.S. Savas is professor of public systems management and director of the Center for Government Studies in the Graduate School of Business of Columbia University. He is also associate director of the Center for Policy Research. Formerly he was First Deputy Administrator of the City of New York. He has served as a consultant to local governments and the federal government in the area of productivity. Professor Savas is well known for his research and writings on the role of the private sector in delivering public services, and on the effect of competition on productivity in such services.